# Intelligent Sensor Systems

**Sensors Series**

Series Editor: **B E Jones**

Other books in the series

**Current Advances in Sensors**
Edited by B E Jones

**Solid State Gas Sensors**
Edited by P T Moseley and B C Tofield

**Techniques and Mechanisms in Gas Sensing**
Edited by P T Moseley, J O W Norris and D E Williams

**Hall Effect Devices**
R S Popovic

**Sensors: Technology Systems and Applications**
Edited by K T V Grattan

**Thin Film Resistive Sensors**
Edited by P Ciureanu and S Middelhoek

**Biosensors: Microelectrochemical Devices**
M Lambrechts and W Sansen

**Sensors VI Technology, Systems and Applications**
Edited by K T V Grattan

# Intelligent Sensor Systems

John Brignell and Neil White
University of Southampton

Institute of Physics Publishing
Bristol and Philadelphia

© IOP Publishing Ltd 1994

All rights reserved. No part of this publication may be reproduced, stored in a retrieval system or transmitted in any form or by any means, electronic, mechanical, photocopying, recording or otherwise, without the prior permission of the publisher. Multiple copying is only permitted in accordance with the terms of licences issued by the Copyright Licensing Agency under the terms of its agreement with the Committee of Vice-Chancellors and Principals.

British Library Cataloguing in Publication Data

A catalogue record for this book is available from the British Library

ISBN 0 7503 0297 6

Library of Congress Cataloging-in-Publication Data are available

Series Editor: **Professor B E Jones,** Brunel University

Published by Institute of Physics Publishing, wholly owned by
The Institute of Physics, London
Institute of Physics Publishing, Techno House, Redcliffe Way, Bristol BS1 6NX, UK

US Editorial Office: Institute of Physics Publishing, The Public Ledger Building, Suite 1035, Independence Square, Philadelphia, PA 19106, USA

Typeset by the authors

Printed in the UK by Bookcraft, Bath

# Contents

|   |   |   |
|---|---|---|
| **Preface** | | x |
| **1** | **Basics** | **1** |
| | 1.0 Introduction | 1 |
| | 1.1 What is Measurement? | 1 |
| | 1.2 What is Information? | 4 |
| | 1.3 Information Flow in Measurement Systems | 6 |
| **2** | **Signals and Systems** | **9** |
| | 2.0 Introduction | 9 |
| | 2.1 A Survey of Classical Systems Theory | 9 |
| |     2.1.1 System description | 9 |
| |     2.1.2 Response | 12 |
| |     2.1.3 Differential equations | 12 |
| |     2.1.4 Time and frequency domains | 13 |
| |     2.1.5 The fundamental theorem of algebra | 14 |
| |     2.1.6 Poles and zeros | 15 |
| |     2.1.7 Convolution and deconvolution | 17 |
| |     2.1.8 Signals as stochastic processes | 18 |
| | 2.2 Sampled Signals and Discrete Systems | 22 |
| |     2.2.0 Introduction | 22 |
| |     2.2.1 The sampling process | 23 |
| |     2.2.2 Hold | 24 |
| |     2.2.3 The $z$-transform | 24 |
| |     2.2.4 Sampling Theorem | 26 |
| |     2.2.5 A simple process | 27 |
| |     2.2.6 First and second order systems | 30 |
| |     2.2.7 Root loci | 31 |
| **3** | **Physical Principles of Sensing** | **33** |
| | 3.1 General Principles of Transduction | 33 |
| |     3.1.1 Physical variables | 34 |
| | 3.2 Primary Sensor Defects and Their Compensation | 36 |

|   |     |                                                          |     |
|---|-----|----------------------------------------------------------|-----|
|   | 3.3 | Survey of Primary Sensing Mechanisms                     | 42  |
|   |     | 3.3.1 Mechanical                                         | 42  |
|   |     | 3.3.2 Thermal                                            | 65  |
|   |     | 3.3.3 Chemical                                           | 71  |
|   |     | 3.3.4 Magnetic                                           | 78  |
|   |     | 3.3.5 Radiant                                            | 82  |

# 4 Electronic Measurement Techniques — 88

|     |                                                          |     |
|-----|----------------------------------------------------------|-----|
| 4.1 | Transducer Interface Circuits                            | 88  |
|     | 4.1.1 Bridge circuits                                    | 88  |
|     | 4.1.2 Non-linear bridge elements                         | 91  |
|     | 4.1.3. Low-power interfacing                             | 92  |
| 4.2 | Operational Amplifiers                                   | 93  |
|     | 4.2.1 Instrumentation amplifier                          | 94  |
|     | 4.2.2 High performance amplifiers                        | 95  |
|     | 4.2.3 Isolation amplifier                                | 96  |
|     | 4.2.4 Logarithmic amplifier                              | 97  |
|     | 4.2.5 Charge amplifier                                   | 99  |
| 4.3 | Data Conversion                                          | 100 |
|     | 4.3.1 Digital-to-analogue converters (DACs)              | 100 |
|     | 4.3.2 Analogue-to-digital converters (ADCs)              | 102 |
|     | 4.3.3 Gain control                                       | 109 |
| 4.4 | The Effect of Noise on Analogue Systems                  | 110 |
|     | 4.4.1 Thermal noise                                      | 110 |
|     | 4.4.2 Shot noise                                         | 111 |
|     | 4.4.3 $1/f$ noise                                        | 111 |
|     | 4.4.4 Interference, screening, shielding, and grounding  | 112 |
|     | 4.4.5 Recovering the signal                              | 112 |
| 4.5 | Quantization and Other Digital Noise Sources             | 115 |
|     | 4.5.0 Introduction                                       | 115 |
|     | 4.5.1 Fundamental non-linearity of quantization          | 115 |
|     | 4.5.2 Treatment as a noise problem                       | 116 |
|     | 4.5.3 Importance of word length                          | 117 |
|     | 4.5.4 Possibility of amplification of noise in digital systems | 118 |
|     | 4.5.5 Truncation and rounding                            | 120 |

# 5 Enabling Technologies — 122

|     |                                |     |
|-----|--------------------------------|-----|
| 5.0 | Introduction                   | 122 |
| 5.1 | Traditional Techniques         | 122 |
| 5.2 | Silicon Planar Technology      | 125 |
|     | 5.2.1 Introduction             | 125 |

|  |  | 5.2.2 The silicon planar process | 126 |
|---|---|---|---|
|  |  | 5.2.3 Silicon micromachining | 129 |
|  | 5.3 | Thin-film Technology | 131 |
|  |  | 5.3.1 Evaporation | 131 |
|  |  | 5.3.2 Chemical vapour deposition | 132 |
|  |  | 5.3.3 Sputtering | 133 |
|  |  | 5.3.4 Langmuir-Blodgett films | 134 |
|  | 5.4 | Thick-film Technology | 134 |
|  |  | 5.4.1 The thick-film production process | 135 |
|  |  | 5.4.2 Thick-film sensors | 136 |
|  | 5.5 | Optical Fibre Technology | 137 |
|  |  | 5.5.1 Intensity Modulation | 138 |
|  |  | 5.5.2 Phase Modulation | 138 |
|  |  | 5.5.3 Frequency Modulation | 140 |
|  |  | 5.5.4 Polarization Modulation | 140 |
| 6 | **Intelligent Sensor Concepts** |  | **142** |
|  | 6.0 | Introduction | 142 |
|  | 6.1 | Elements of Intelligent Sensors | 143 |
|  | 6.2 | Structures | 146 |
|  |  | 6.2.1 Hardware structures | 146 |
|  |  | 6.2.2 Software structures | 152 |
|  | 6.3 | Processes And Procedures | 159 |
|  |  | 6.3.0 Introduction | 159 |
|  |  | 6.3.1 Digital filters | 159 |
|  |  | 6.3.2 Differentiation and integration | 163 |
|  |  | 6.3.3 Smoothing | 166 |
|  |  | 6.3.4 Discrete transformation | 166 |
|  | 6.4 | Applications | 176 |
|  |  | 6.4.0 Introduction | 176 |
|  |  | 6.4.1 Petit mal detector | 176 |
|  |  | 6.4.2 Sensor array processing | 177 |
|  |  | 6.4.3 Sensor response compensation: the load cell as an example | 179 |
|  |  | 6.4.4 The intelligent building | 190 |
|  |  | 6.4.5 Automotive applications | 193 |
|  | 6.5 | Production | 197 |
|  |  | 6.5.0 Introduction | 197 |
|  |  | 6.5.1 Production costs | 198 |
|  |  | 6.5.2 Trimming | 198 |
|  |  | 6.5.3 Intelligent sensor trimming | 199 |
|  | 6.6 | Stimulus-response Sensors (Unorthodox Stimuli) | 200 |
|  | 6.7 | Discussion | 203 |

| | | |
|---|---|---|
| **7** | **Communication And Sensor Networks** | **204** |
| 7.1 | System Topologies | 204 |
| | 7.1.1 Prioritization | 205 |
| 7.2 | General Requirements of a Low Level Protocol | 207 |
| 7.3 | Protocol Implementations | 209 |
| | 7.3.1 Communication systems reference model | 209 |
| | 7.3.1 Conventional instrumentation communication systems | 211 |
| | 7.3.3 Manufacturing Automation Protocol (MAP) | 212 |
| | 7.3.4 Enhanced Performance Architecture (EPA) MAP | 213 |
| | 7.3.5 Fieldbus | 213 |
| | 7.3.6 The HART communication protocol | 214 |
| | | |
| **8** | **Physical Realizations** | **216** |
| 8.0 | Introduction | 216 |
| 8.1 | A Magnetic Field Sensor | 217 |
| | 8.1.0 Introduction | 217 |
| | 8.1.1 Mechanisms | 217 |
| | 8.1.2 Structural compensation | 218 |
| | 8.1.3 Monitored compensation | 219 |
| | 8.1.4 Information processing | 219 |
| | 8.1.5 Tailored compensation | 219 |
| | 8.1.6 Discussion | 219 |
| 8.2 | The Intelligent Load Cell | 220 |
| | 8.2.0 Introduction | 220 |
| | 8.2.1 Mechanisms | 221 |
| | 8.2.2 Structural compensation | 222 |
| | 8.2.3 Monitored compensation | 225 |
| | 8.2.4 Information processing | 225 |
| | 8.2.5 Discussion | 227 |
| 8.3 | An Optically Coupled Pressure Sensor | 227 |
| | 8.3.0 Introduction | 227 |
| | 8.3.1 Mechanisms | 228 |
| | 8.3.2 Monitored compensation | 229 |
| | 8.3.3 Information processing | 229 |
| | 8.3.4 Discussion | 229 |
| 8.4 | The Electronic Nose | 231 |
| | 8.4.0 Introduction | 231 |
| | 8.4.1 Mechanisms | 231 |
| | 8.4.1 Structural compensation | 232 |
| | 8.4.2 Monitored compensation | 232 |
| | 8.4.3 Information processing | 232 |
| | 8.4.4 Discussion | 235 |

| | | |
|---|---|---|
| 8.5 | ASICs | 235 |
| | 8.5.0 Introduction | 235 |
| | 8.5.1 A strategy for intelligent sensor ASICs | 238 |
| | 8.5.2 Discussion | 242 |

**9 Conclusion** — 243

**References** — 245

**Index** — 249

# Preface

This volume, one in an ongoing series on sensors, is, like its predecessors, a research monograph. Unlike many such works, however, it does contain a certain amount of introductory material, for reasons we attempt to justify below. The sensor field is in itself a multi-disciplinary one, but the particular aspect we are concerned with here is even more so. As such, the treatment of the subject has set the authors a number of posers which proved difficult to solve, and in order to deal with the sheer amount of potential material many difficult compromises were reached. How successful these have been can only be left to the judgement of the reader. The intended readership is product innovators in industry and academic researchers, together with undergraduate and postgraduate students who are specializing in instrumentation.

Producing a text-book in the field of applied electronics is in many ways a thankless task. There are the usual problems of authorship:

Words strain, crack and sometimes, break under the burden,
Under the tension, slip, slide, perish,
Decay with imprecision, will not stay in place,
Will not stay still .......
(Burnt Norton)

In this technology, however, there is the additional problem that the words are out of date almost as soon as they hit the magnetic oxide, let alone the printed page. Because writing is essentially a slow and time-consuming process the result is like a snapshot of a moving object taken with a slow shutter speed. One hopes that there is enough information and interest in the resulting blur to have made the exercise worthwhile.

The IOPP series on sensors has, hitherto, valuably and comprehensively covered a wide range of basic sensor mechanisms. This volume is different in a number of ways. First, its title bears the word *intelligent*, at best an unsatisfactory locution. We have on occasions flirted in research publications with the American preferred alternative *smart* but in the UK this tends to have a slightly pejorative sense, especially when followed by *Alec* or a more fundamental colloquialism; though in the light of some of the forewarnings we give later in this text such a spirit might not be entirely inapposite.

The second difference in the title is the appearance of the word *system*. In our context this has a double meaning. The sensor itself may be regarded as a system, and we shall leave the definition and implications of that term until later in the text, while the sensor also may be regarded as a sub-system of a larger host system, a concept with profound implications, some of which we explore in the following pages.

The third notable difference is that books in this series have dwelt, with advantages, on the details of the science and technology of particular sensing mechanisms and their applications. Here we have rather concentrated on the global rules of the game, tried to give an account of the flavour of the subject, yet have peppered that account with sharp reminders of some fundamental and very pertinent hazards: for this is a field in which 'a little learning is a dangerous thing' and to continue the quotation beyond the conventional cliché urge the reader to 'dig deep, or taste not the Pierian spring'. Many of the rules and admonitions in this text arise from actual observation of students, both undergraduate and postgraduate, but also of practising engineers and scientists, attempting to apply 'common-sense' solutions which are flawed by a lack of appreciation of the fundamentals, sometimes to the point of disaster.

In preparing this text we agonized at length over what to put in and what to leave out. The experience of presenting this sort of material to various audiences at lectures and seminars has revealed that people find some of it highly informative and some of it trivial. Unfortunately, different people put different bits into each category. In response we have tried to compromise by giving a treatment to most parts of this multi-disciplinary area. Naturally it is somewhat terse in parts, but at least it gives signposts to the things that are important. The alternative would be to produce a multi-volume encyclopaedia, which would be beyond the endurance of the authors and the tolerance of the reader. We would anticipate that most readers will find many passages that are familiar enough to skip, but hope that they will find it useful to have a reminder of the fundamental building blocks of the subject. Furthermore, perhaps surprisingly, we were somewhat relieved when a review of intelligent sensor technology appeared when we were in the late stages of preparing the first draft of this text (Ohba 1992). That text concentrated very much on the practicalities of the topic, with many photographs and detailed drawings of actual devices and systems. Thus we felt free to concentrate on the areas where there were substantial lacunae in the literature, particularly the fundamentals, areas where mistakes could be costly – indeed, areas representing the very basic philosophy of the subject.

When one of the authors was writing the introduction to an earlier book (Brignell and Rhodes 1975), two decades before this one, he was obliged to include an apologia for the very idea of using computers in measurement; for in those days measurement science was dominated by the high priests of the discipline, the keepers of the mysteries, who, by their own account, could obtain phenomenal accuracies by true observance of the rites. To these practitioners the computer was a frivolous intrusion. Talk of subjectivity in human measurement was a heresy to be suppressed.

Yet it was the very observation of such subjectivity that had turned that author to the use of the computer in measurement in the first place. The particular field of activity at that time was the measurement of electric breakdown strength. A simple yet rigorous statistical argument predicted that such measurements should be highly scattered with a characteristic skewed (extreme-value) distribution; yet the measurement specialists were reporting higher and higher values, with smaller and smaller scatters of normal distribution. They were searching for a holy grail known as 'Intrinsic breakdown strength', now accepted as a chimera, but then verity.

It was only by remotely recording the efforts of some student volunteers to repeat such measurements that it was demonstrated how they were achieved. It was vaguely indicated to the volunteers that high and consistent results were somehow laudable, before they were given a knob to control a high voltage supply and an electrostatic voltmeter to observe the voltage reached before breakdown. The results were remarkably consistent. Each volunteer would start by applying the voltage as a slow linear ramp, but after a few shots they began to learn where the breakdown was likely to occur, and they would speed up the rate of rise at first then slow it down as the expected value was reached. The early linear ramps were gradually transformed into curves which approached the well known $1-e^{-t}$ shape. Now, because the breakdown event is random in time as well as voltage, what they were doing was giving breakdown more time to occur where it was expected and therefore effectively forcing it to occur there. Furthermore, there were just a few 'freak' low values, which by the $3\,\sigma$ rule of thumb for normal distributions could be discarded, as had undoubtedly been done in the past. The same experiments repeated by means of electronic ramp generation and data recording produced results entirely in line with the predictions of the simple statistical theory based on the idea of a chain being as strong as its weakest link.

It behoves us, however, not to become too complacent. Computer programs are also written by human beings, and therefore contain their prejudices, lacunae of knowledge, lapses of logic and sheer blunders. In the two decades since that earlier text particular errors and blind-spots have been repeated over and over again in the experience of the authors, attempts to smooth data being a classic example, which we shall take up later in these pages. It is very easy to think up a glib stratagem and write a line or two of program to modify a signal, but actually understanding what that process really does is a different matter, and that is a major theme of this text. The human sub-conscious is always at work. It has its own biases and expectations, and has subtle ways of bringing them into play, without the knowledge of the conscious mind. The price of objectivity in measurement is eternal vigilance.

The development of sensors in nature was a major factor in the development of the higher species. The appearance of a crude phototropism in a primitive cell created a small advantage, which by the inexorable process of survival of the fittest, led to the eye of the eagle. The emergence of a rudimentary chemical sensitivity led to the nose of the bloodhound. Yet man, whose organs of sense can match neither of these examples, became the dominant species. Why?

The reason is, of course, that in addition to his modest organs of sense, man developed his great brain to process the information they supplied, and was thus able to conquer his environment. Equally there is a process of technological evolution in which the fittest survive, fitness being determined by price, performance and subtler factors such as social acceptability and man's war-like tendencies.

We are concerned here with the state of evolution in which processing power is coming to the aid of the sensing process, but there is also a dramatic divergence from nature, in that we are proposing a development in which each sense organ has its own little brain.

Before leaving the biological analogies let us consider one other factor. Man is more likely than other creatures to be misled by optical illusions. This is because the great leaps of deduction he makes on modest sensor information can deceive when he comes up against artificial constructs that are not generally found in nature; but he is not unique in this. The trout rising to an artificial fly has fatally mistaken a bunch of feathers for a tasty morsel. There are analogous mistakes that can be made by the designer of intelligent sensor systems. He can be tempted to build a tower of deduction on weak foundations of information, with the inevitable result. Apparently simple ideas like quantization, differentiation and smoothing can be serious hazards if not properly understood. In this text with its bibliography and references, which we have striven to keep to an essential minimum, we seek to inform the reader of the enormous power of intelligent sensor systems, while reminding him of the care that has to be taken with the fundamental principles.

It is easy to understate the profound change wrought by the introduction of digital processing into sensors. It is more than just another small step in the evolution of a class of devices important to civilization. The inevitable defects of devices that we have become accustomed to live with now become invisible, as they can be dealt with in the sensor housing. The mathematical methods developed by the giants of the past ( such as Oliver Heaviside) to enable us to understand how systems operate now form the basis of processes and procedures which become essential parts of the devices. Of the enormous range of potential physical mechanisms for sensing most had to be rejected because of drawbacks that can now be overcome. Engineers have been inclined to make do with measurements that they think they can make, rather than those that they really need to make. All these observations call for a new way of thinking by the designers, the manufacturers and the users of sensors. There is a great liberation for system designers, who now have the promise of being able to get on with their proper job and not spend time trying to cope with the defective behaviour of the devices available to them. This process of 'compartmentalization' is one of the marked benefits of the technology, since failing to see the wood for the trees has been one of the hazards of complex system design. In freeing the system designer in this way, the sensor designer has to widen his horizons, become more of a polymath, and it is the object of these pages to present a travellers' guide to this new land. As such, we are painfully aware that the need for brevity makes this treatment all too sketchy, but hope that it will at least point some readers in the right direction.

When one of the authors first adumbrated some of the ideas leading to intelligent sensors at a conference at the beginning of the 1980s a well known industrialist was heard to say in a loud stage whisper 'Pie in the sky!', and there were one or two 'Hear, hears'. The reason for that response was that at that time a microprocessor cost considerably more than a basic sensor and it did not make sense to dedicate the former to the latter. What this criticism did not take account of was the trends in cost and performance of microelectronics. The question of trends is one we return to later in this text, and is one that should pre-occupy researchers and developers in the field. It is interesting that the said industrialist's company now boasts intelligent sensors in its range of products. A decade is a long time in electronics, and it is hoped that this text will help those who are planning for the future.

It is important to state at the outset what we mean by intelligent sensors, as the usage is widely variable. Some authors have applied the term to sensors that have a few buffer amplifiers on-chip. In this text we restrict the term to sensors which have at least one digital processor. Since such a processor is now simply a library element in the repertoire of chip sub-systems, this is no longer the extreme view it might have been considered by some at the time we set out on preparing this account.

We begin this text by addressing in Chapter 1 the simple yet profound questions of what we mean by measurement and information. In Chapter 2 we outline the essential results of signal and system theory needed for an understanding of intelligent sensor processes, with particular emphasis on the consequences of combining continuous and discrete sub-systems. Chapter 3 is concerned with the physical principles on which sensors are based and how they are exploited in the construction of some of the more important primary sensor elements. In Chapter 4 some of the essential electronic measurement techniques are reviewed in the light of the requirements of intelligent sensors.

An important aspect is the range of enabling technologies that can be utilized in the realization of the devices, and these are described in Chapter 5. Chapter 6 is concerned more with the concepts which are unique to the intelligent sensor approach to measurement. This chapter is in many ways the keystone of the arch. One of the prominent characteristics of intelligent sensors is the efficient way they are able to communicate with a host system, and this is summarized in Chapter 7. Finally in Chapter 8 we review some of the current physical realizations in the form of case studies and in Chapter 9 draw a few conclusions.

It is a dreadful cliché for authors to dedicate their works to their spouses, but anyone who has been involved in such an exercise will understand why we dedicate this text to Gillian and Sylvia with thanks for their patience and tolerance.

John Brignell
Neil White
June 1993

# 1

# Basics

## 1.0 INTRODUCTION

Sensor technology is concerned with two activities whose history goes back to the dawn of civilization, **measurement** and **information processing**. Measurement is a process which is a necessary, though not sufficient, condition for the existence of civilization. In early Egypt it was only when people developed the means of measuring land areas and water volumes that they could organize agriculture and the necessary irrigation. Later they were able to develop a means of processing the numbers produced by measurement, i.e. mathematics, which brought a new dimension to the ways that those numbers could be exploited. Of course information processing had first been developed by nature in the form of the animal brain, which reached its zenith with the human form. Developments in nature only occur if they provide an evolutionary advantage, i.e. an increase in fitness to survive, and it is probable that man's great brain developed to enable him to knock hell out of the tribe next door, and by subsequent activities pass on his genetic advantage to future generations. It is an enduring irony that one of the greatest impetuses to sensor development has been man's aggression against his own kind.

## 1.1 WHAT IS MEASUREMENT?

Measurement is a process of assigning numbers to entities and events in the real world. In the case of events the numbers usually represent time, a simple exception being the case where the measurement process is one of counting events. Indeed, this exception is important in that it is the only example of scientific measurement which can produce a unique number as a result. The general rule is that a scientific measurement maps an event or entity onto a range of numbers. Thus the statement 'Jim is six feet tall' is not valid scientifically; whereas the statement 'Jim is between five foot eleven inches and six feet one inch tall' is a valid one.

In the more formal terms of the axiomatic probability theory, a study of which is essential to the measurement scientist (Papoulis 1965), we conceive of a measurand deriving from a population, such as the heights of men, which is described by a probability density function. The probability density function for a random variable $x$ is defined as

$$f(x) = \lim_{x \to 0} \frac{Prob\{x \le \pmb{x} \le x + dx\}}{dx} \qquad (1.1)$$

and if $f(x)$ is continuous it follows that

$$Prob\{\pmb{x} = x\} = 0. \qquad (1.2)$$

The precision of any measurement process is bounded, that is to say there is a range of numbers $\Delta x$, which cannot be distinguished from each other by the process. A simple example is the range of numbers represented by the Least Significant Bit (LSB) of an analogue-to-digital converter (ADC). A more complicated example is the noise level in an amplifier, which makes $\Delta x$ itself a stochastic variable. This means that at level $x$ a proportion of the parent population $f(x) \Delta x$ cannot be distinguished from each other. Thus to continue the above example, if the purpose of the measurement of height is to identify a criminal a target height of three foot eight will narrow down the search much more effectively than one of five foot eight, owing to the differences in the value of $f(x)$ for the human population, whatever the precision $\Delta x$. An important outcome of intelligent measurement is the ability to classify entities or events via one or more measures or metrics without human intervention.

It is important to distinguish carefully between precision and accuracy. A measurement made to a precision of 16-bits or one mV is not necessarily one made to an accuracy of 16-bits or one mV. Imprecision is just one component of inaccuracy, and other components, which in a particular circumstance may be more dominant, comprise a large number of potential measurement errors, including, in the case of intelligent measurement, errors arising in the processing of the numbers. This is not a trivial consideration in intelligent measurement systems, as there is always a temptation to build up a structure which is too ponderous for the foundations of accuracy on which it is based, and it is not always easy to trace the flow of inaccuracies through a complex digital processing structure.

Restrictions of space limit the discussion here on measurement errors, but this is not to play down their importance, which is crucial to the theme of this text, and if in doubt the reader should acquaint himself with the fundamentals of the classification and treatment of errors (Jones 1980, Collett and Hope 1983, Doebelin 1990).

A single act of measurement produces a number. In itself this number has no worth until it is qualified by some interpretive knowledge. Returning to our simple example, let us create a progression of enhanced worth

six
**six feet**
**six feet tall**
**Jim is six feet tall**
**Jim is six feet tall to an accuracy of plus or minus one inch.**

In an instrumentation computer the result of this measurement would be a simple binary number (110), and the qualifications which increase the worth of this number would not normally appear in the computer program (unless it is part of a particularly sophisticated knowledge based system, which is outside the scope of the present treatment). The qualifiers are useful to the user or programmer of the system in interpreting the number, and narrowing down the range of entities that it can represent. By examining the functions of each of the additions in the above progression we can derive a breakdown of the component parts of the result of a single act of measurement.

**Six** is the fundamental quantitative outcome of the measurement, the **number**.

**Feet** is the **unit** of measurement and it tells us two things, the **dimensionality** of the measurand (length) and the **system** of units (imperial).

**Tall** acts as the **specifier** of the measurand, i.e. it is height rather than depth, width, wavelength or indeed anything else that can be measured in the units.

**Jim** is the **origin** of the measurand or in instrumentation terms the identifier of the test object.

**One inch** is the accuracy of the measurement ( of which precision is a subset ).

Thus our complete measurement comprises the following parts

### Number, Unit, Specifier, Origin, Accuracy,

and each contributes to the worth of the measurement in human terms, though not necessarily in computer terms. In a closed-loop control system for example only the **Number** would be of relevance.

There are many important aspects of measurement which cannot be treated here (standards, traceablity, etc.). Highly accurate measurement is a specialist field in its own right, but on the whole it is not the concern of intelligent measurement systems, which make their greatest contribution where the question is often whether a measurement can be made at all rather than whether it can be made to an accuracy of, say, one part in $10^8$.

To summarize, the process of measurement produces numbers which are in correspondence with sets of entities or events in the real world. The worth of those numbers in human terms is enhanced by the addition of qualifiers, each of which narrows down the incertitude in our knowledge of the world.

## 1.2 WHAT IS INFORMATION?

**Information** – intelligence given; knowledge: an accusation given to a magistrate or court.  *Chambers Twentieth Century Dictionary*

One of the great problems for scientists, particularly those working in a rich language such as English, is the imprecision and associations carried by the everyday words that we have to use. **Information** is a case in point. We need a working definition which is simple and free of contextual associations. The following will serve

**Information** - that which decreases uncertainty.

This, of course, begs the question of what we mean by uncertainty. Consider arrival at a Y junction in the road. If there is no sign post we are in a state of total uncertainty. In this case the basic unit of information, a sign post with two fingers correctly labelled, will remove that uncertainty. If the roads branch similarly further on, similar signposts will give similar amounts of further information. At this stage, whichever route we take, two signposts enable us to select one of four possible paths. A third set of signposts on those four paths enable us to select one of eight possible paths. Evidently after passing $I$ signposts we have selected one out of $2^I$ possible paths. Thus $I$ binary units of information (or bits) will enable us to distinguish between $N = 2^I$ possibilities. Thus our basic unit of information is the **bit**, which is generally grouped into entities known as **words**, the special term **byte** being reserved for a word of eight bits. In the measurement of volume we use the same units, litres, to denote the capacity of a vessel and its contents: the same goes for computer words and it is important not to confuse the capacity with the contents. Thus the length of the word in bits represents its **maximum** information content, and in general an $n$-bit word will contain less than $n$ bits of information.

It is not possible here to go into detail of the treatment of information in engineering terms, and the reader is urged to familiarize himself with the main results of information theory (Goodyear 1971). However, a few general remarks are relevant. Information theory can take little account of contextual factors. Nevertheless, given the prior information for example, that our bytes represent letters of the English language (e.g. ASCII encoding), we can make certain statements. The letter X provides much more information than the letter E, as any crossword puzzler will confirm, as it is less probable and therefore removes more uncertainty. However, information theory is unable to take account of higher contextual factors, such as the cultural experience. Hence an arbitrary group of words such as 'The gostak distims the doshes' (Conklin 1962) or 'God save the Queen' would have equal status in information theory and would be evaluated purely in terms of the probabilities of the letters involved. An Englishman might claim that adding the last letter to *God save the Quee_* adds no information, though it might be an '*r*' which, as it happens, is equally probable with an '*n*'; but this is a purely cultural distinction. Certainly the '*u*' after the '*Q*' offers no information in English since it arrives with unit probability.

Obviously, at the present stage of instrumentation we are not greatly concerned with the English language as an information source, but it is useful to persist with it briefly as a model.

Any discrete source of information will produce symbols from a restricted alphabet, 256 different numbers from an 8-bit ADC or 26 letters plus the space in English. If the symbols were equiprobable, then the quantity of information per symbol would be simply given by the inversion of the above relationship:

$$I = \log_2(N) = \log_2(1/p) = -\log_2(p) \quad \text{bits} . \qquad (1.3)$$

On this basis English text would convey $\log_2(27)$ or 4.75 bits of information per symbol. Of course the letters are not equally probable, so an important quantity is the average amount of information per symbol. To find the average we have to multiply the probability of each symbol by the quantity of information it carries and sum over all the symbols.

$$H = \sum_j p_j \log_2(1/p_j) = -\sum_j p_j \log_2(p_j) . \qquad (1.4)$$

This important quantity is known as the source **entropy**, which is analogous, with entropy in thermodynamics and in general is a measure of disorder. A long message of $n$ symbols from a source of entropy $H$ will contain $nH$ bits of information. Taking account of the different probabilities of letters of the alphabet and also the conditional probabilities associated with their pairings English text has a source entropy of 3.3 bits/symbol rather than the ideal 4.75 above.

In the case of our ADC the idealistic source entropy would be 8-bits, but we have to choose the pre-amplification to minimize the probability of the input signal going out of range and the probability of large or small numbers is much less than the probability of middle range numbers, so the actual source entropy is much less than 8-bits per conversion. The actual rate of information coming from the converter is controlled by the distribution and the bandwidth of the signal, since these control the source entropy and the sampling rate. This is a very important consideration in the treatment of binary buses, which will figure prominently later in this text. Two important related quantities in the encoding of information are **efficiency** and **redundancy**. If a code of length $L$ is used to represent symbols from a source of entropy $H$ then the efficiency is given by

$$\eta = H/L \qquad (1.5)$$

and the redundancy is given by $1 - \eta$. Thus a five bit code to transmit upper case English text without punctuation would have an efficiency of 0.66 and a redundancy of 0.34. In message transmission much higher redundancies are used to secure the integrity of transmission, and are achieved by the addition of parity or higher order checks.

English text is itself about 50 % redundant, so it is possible to understand quite grossly corrupted messages, e.g. 'Gid suve thx Qyeen'.

We emphasize that information theory has nothing to say about the qualitative aspects of information, e.g. truth, value or sensibility. This applies not only to English text but also to signals from our ADC, and in both cases what is transmitted may well be nonsense in terms of human interpretation, but is of information theoretic equal value to a signal which conveys humanly meaningful data.

One of the intelligent sensor functions we shall have to deal with later in this text is the question of data condensation. In information theory terms this means reduction of flow rate of entropy. For the latter, as in the thermodynamic case, it is implied that the process involved is irreversible, and only a reversible process can maintain the entropy constant. Furthermore, since it is not in the natural order of things for entropy to decrease it is implied that work (or in our case computation) has been done on the fluid (information).

In summary, information to us is a quantity which is identifiable and measurable, just like other scientific quantities, such as magnetic flux or heat. An instrumentation system is one in which this quantity flows and is subject to rules of continuity as are physical quantities. This point we pursue in the following section.

## 1.3 INFORMATION FLOW IN MEASUREMENT SYSTEMS

As we have seen, information, at least in the engineering sense, is a quantity which can be identified and measured and located. As such it can be treated as any other physical quantity such as electric field, heat or fluid, and from a simple consideration of what can happen to such a quantity relative to an enclosed volume we can, as with those quantities arrive at some basic but powerful design rules.

If we consider an arbitrary quantity within a closed surface there are only six different things that can happen to that quantity in a given time interval. This is illustrated in figure 1.1, in which we can see that there can be a quantity input, $Q_I$, a quantity output, $Q_O$, stored, $Q_S$, unstored, $Q_U$, destroyed, $Q_D$ or created, $Q_C$. As we have specified that these are the only possible things which can occur to the quantity it follows that the algebraic sum of all these six items must remain constant and furthermore the sum of their rates of change must be zero. In any particular case some of these quantities are zero; thus for magnetic fields $Q_C$ and $Q_D$ are zero while for electric fields they are determined by the charge within the surface, yet for incompressible fluids $Q_S$ and $Q_U$ are zero but finite in the compressible case. In each case by taking into account all the terms of the equation we come to a fundamental physical law ( e.g. Gauss).

# BASICS

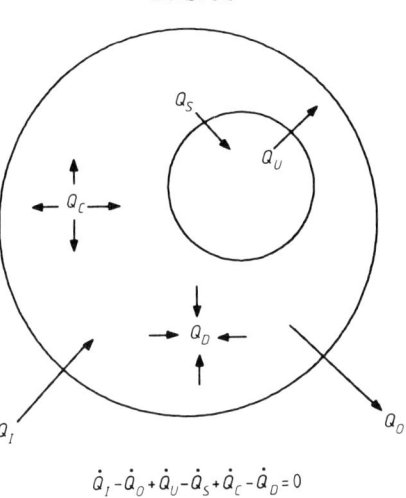

[Figure: concentric circles with flow arrows labeled $Q_S$, $Q_U$, $Q_C$, $Q_D$, $Q_I$, $Q_O$]

$$\dot{Q}_I - \dot{Q}_O + \dot{Q}_U - \dot{Q}_S + \dot{Q}_C - \dot{Q}_D = 0$$

**Figure 1.1** Consideration of all that can happen to any quantity with respect to a closed surface leads to a general continuity equation.

Now applying this fundamental and simple idea to information as defined in the previous section we see that all of the six things may occur. The only one which might cause us difficulty is $Q_C$.

However, the question of the creation of information is a deeply philosophical one, and for our purposes it is safe to assume that information cannot be created within the systems we are describing, i.e. it is something that enters from the outside. It **can**, however, be destroyed. The deliberate discarding of information is a normal function of intelligent sensors, and will be discussed later in terms of data condensation (§ 6.1). The inadvertent loss of information, on the other hand, can be a serious matter.

An example of information loss is the case of buffer overflow. In order to cope with variations of input and output data rates of our intelligent sub-system we may have to provide a buffer store.

The continuity equation of figure 1.1 in these circumstances becomes the basic design equation to determine the size of buffer required. In order to utilize the equation, however, we have to have models for the way the quantities behave. $Q_I$ for example, might be determined by a constant sampling rate or by some externally controlled demand cycle; while $Q_O$ could be controlled by a constant polling cycle or, more complicatedly, by the essentially stochastic process of gaining access to a shared bus. Thus in the general case, whatever value we chose for the size of buffer store, there is a finite probability that

$$Q_I - Q_O + Q_U - Q_S \quad (1.6)$$

will become positive, meaning that $Q_D$ must become positive, and information **must** be lost. This area of design is not an easy one to encapsulate, but it is important. If we must perforce use relatively crude models for the behaviour of the external sources and sinks of information then we are obliged to make pessimistic assumptions, and in general our buffers will be relatively empty.

The above is an important mechanism in forcing all our sub-systems to be dependent upon each other. A sub-system's access to the bus is controlled by the rest of the system's access. Here lies a major design trap; for if the system grows, as distributed systems are inclined to, then the access of every sub-system to the bus is necessarily reduced and the probability of consequent information loss inevitably increased. Needless to say there is a wide variety of malfunction which can have a similarly disastrous effect.

We see that our distributed instrumentation system is analogous to a water supply system (with the flow largely in reverse) and we expect local cisterns to take care of fluctuations of flow; but if the water supply company takes on too many customers or a main bursts then we must expect disruption.

A recurrent theme of this text is that the bus is a precious resource which must be conserved at every stage of hardware and software design.

# 2
# Signals and Systems

## 2.0 INTRODUCTION

The intelligent sensor is an example of a system, and potentially a very complex one. It therefore has to be described in the language of systems theory. Furthermore it is a combination of both continuous and discrete sub-systems. Continuous systems have been around for some time and there is a solid base of theory to describe them, which we might term the classical systems theory. Discrete systems theory is much more recently formulated and may be less familiar to traditionally trained readers. These theories are important to the understanding of intelligent sensor systems and in some areas (such as frequency response compensation) they are absolutely fundamental; so it is necessary to give an account, if all too brief, of them in the context of sensor techniques. If the reader does not feel thoroughly familiar with these concepts he is urged to consult the references (e.g. Papoulis 1965, 1980).

## 2.1 A SURVEY OF CLASSICAL SYSTEMS THEORY

### 2.1.1 System description

A dictionary definition of a system is
**Anything formed of parts placed together or adjusted into a regular and connected whole: a set of things considered as a connected whole; .....a full and connected view of some department of knowledge: an explanatory hypothesis: a scheme of classification: a manner of crystallization: a plan: a method: a method of organization: methodicalness: a systematic treatise.**

Thus a system can be anything from a method of filling in football pools to the human body. At present we are concerned with sensors, the large sub-class of systems which operate on signals. It is important to appreciate that the term signal also has a very wide definition. In our terms, signal is generally taken to mean any quantity which can be represented as a function of time (though the methods of signal processing can often be applied effectively to functions of other variables), and it is usually considered to convey information. The signal function should be single valued, but it may be either continuous or discontinuous. Examples of signals are: the Dow Jones average, EEG potential, air pressure, pH of a solution, etc.

In these terms a system is the mechanism whereby a causal relationship is established between two sets of signals, the input signals ( $x_1, x_2, x_3, ..., x_n$ ) and output ( $y_1, y_2, y_3, ..., y_n$ ) in figure 2.1.

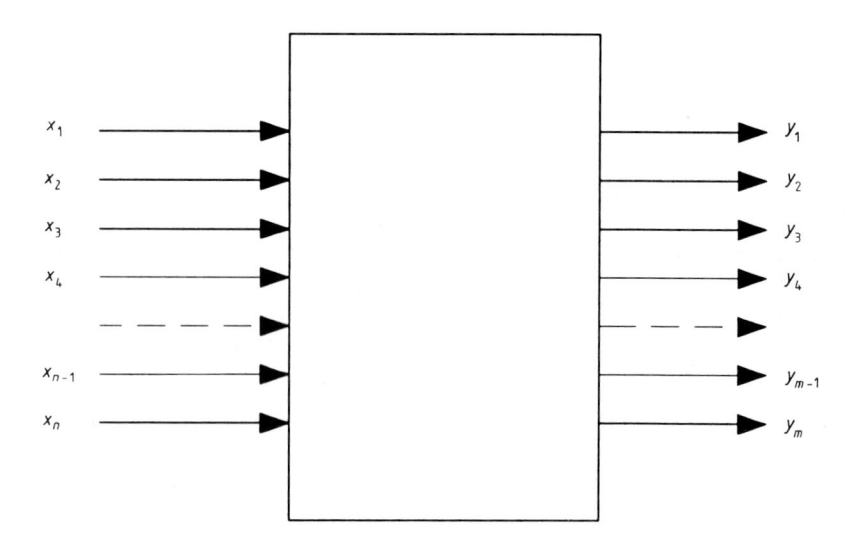

**Figure 2.1** A system as a causal mechanism between inputs and outputs.

The various methods of describing systems are, in effect, methods of describing this causal relationship. The form of description chosen will depend on the nature of the system itself and one's purpose in utilizing or analysing that system. Systems may be classified in various ways, e.g:
    Linear/non-linear
    Random/deterministic
    Continuous/discrete
    Time-varying/unvarying
    With memory/without memory
    etc.

One of the reasons why the mathematical background is so important to an appreciation of the discrete processor at the heart of the intelligent sensor is the fact that it is capable of all the complications which can occur in a system (with the exception that it cannot of itself be random), and this can present both advantages and disadvantages, both of which need to be fully understood by the intelligent sensor systems engineer. Signals may also be classified in various ways, some of which are common with the list of classifications of systems, for example signals may be

    Continuous/discrete
    Bounded/unbounded
    Random/deterministic
    Periodic/aperiodic
    Stationary/non-stationary
    etc.

The most important sub-classification of systems is linear systems. There are two main reasons for this:

1. While no real system is actually linear, most are linear to a good approximation.
2. There is a powerful body of linear algebra which is not matched by a corresponding non-linear form.

Some engineers appear to misunderstand the meaning of the word linear as used in this important context. In particular, it does not mean simply a straight line relationship between output and input signals. In mathematical terms a linear system is one which is additive and homogeneous. That is to say, the result of adding two signals together, or multiplying a signal by constant, is the same whether the operation is performed at the input or the output of the system, figure 2.2.

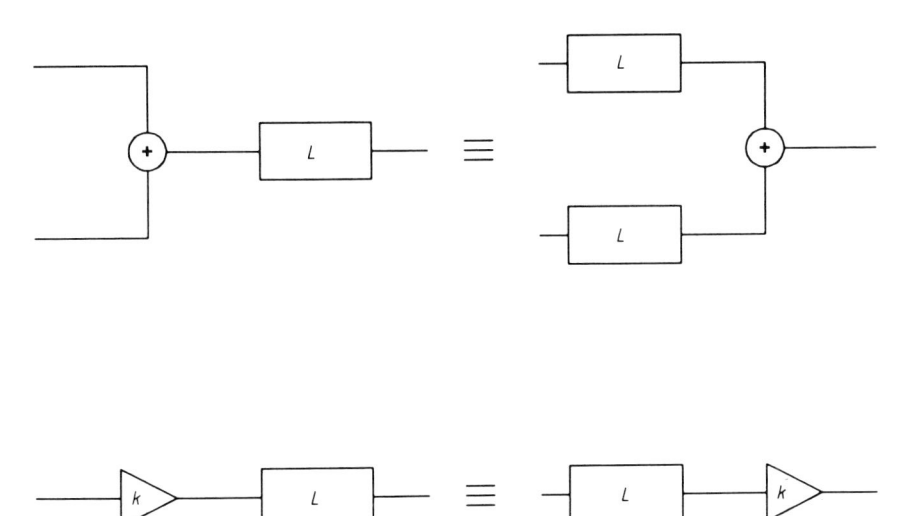

**Figure 2.2** A linear system is additive and homogeneous.

$$\left[\sum a_j \left(\frac{d}{dt}\right)^j\right] x(t) \rightarrow \left[\sum (a_j s^j)\right] X(s) \tag{2.8}$$

and

$$\left[\sum b_i \left(\frac{d}{dt}\right)^i\right] y(t) \rightarrow \left[\sum (b_i s^i)\right] Y(s) \tag{2.9}$$

Hence the Laplace transform converts a difficult problem of the solution of differential equations into the much simpler problem of the solution of polynomial equations. This facilitates the analysis of systems with the aid of known, tabulated Laplace transforms. A few examples of important Laplace transforms are

$$\begin{aligned} \exp(at) &\rightarrow (s-a)^{-1} \\ \cos(at) &\rightarrow \frac{s}{s^2-a^2} \\ \text{step } U(t) &\rightarrow 1/s \\ \text{impulse } \delta(t) &\rightarrow 1 \\ t\exp(at) &\rightarrow (s-a)^{-2} \end{aligned} \tag{2.10}$$

with the important relationships

$$\begin{aligned} \frac{d}{dt} f(t) &\rightarrow s F(s) - f(0) \\ \exp(at) f(t) &\rightarrow F(s-a) \end{aligned} \tag{2.11}$$

### 2.1.5 The fundamental theorem of algebra

This basic theorem has the most important consequences in the theory of both continuous and discrete systems. It comprises the simple statement:

> **Every polynomial of degree *n* has *n* zeros.**

Thus, for example, in the *s*-domain, it is immediately true that the polynomial description of a system above, equations (2.8) and (2.9), may be replaced by the following form

$$\left[\prod (s-s_i)\right] Y(s) = K \left[\prod (s-s_j)\right] X(s) \tag{2.12}$$

# SIGNALS AND SYSTEMS

where $K$ is a constant, and furthermore by re-arrangement

$$Y(s) = K \frac{\prod (s-s_j)}{\prod (s-s_i)} X(s) \qquad (2.13)$$

which means that we may write a system function

$$H(s) = K \frac{\prod (s-s_j)}{\prod (s-s_i)} \qquad (2.14)$$

where

$$Y(s) = H(s) X(s). \qquad (2.15)$$

Thus linear systems may be described as the ratio of two polynomials and also in terms of its poles and zeros, i.e. the roots of the denominator and numerator respectively. A great deal of systems analysis and synthesis may be carried out in terms of these poles and zeros, and we shall use the concept extensively in the discussion of sensor frequency response and its compensation.

## 2.1.6 Poles and zeros

The fact that any linear system may be described, within a constant, by means of the ratio of two polynomials may be exploited to provide a convenient form of geometrical analysis. Thus we have $H$ as a ratio of two polynomials $P$ and $Q$

$$H(s) = \frac{P(s)}{Q(s)} = \frac{P(s)}{\prod (s-s_i)} = \sum \frac{C_i}{(s-s_i)} \qquad (2.16)$$

where for non-repeating factors

$$C_i = \lim_{s \to s_i} \left\{ (s-s_i) H(s) \right\} \qquad (2.17)$$

by l' Hôpital's rule

$$C_i = \frac{P(s_i)}{Q'(s_i)}. \qquad (2.18)$$

So combining

$$H(s) = \sum \frac{P(s_i)}{Q'(s_i)} \frac{1}{(s-s_i)} \qquad (2.19)$$

and inverting

$$h(t) = \sum \frac{P(s_i)}{Q'(s_i)} \exp(s_i t) . \qquad (2.20)$$

This final equation is known as Heaviside's expansion theorem, and the exponential terms are of great significance in systems theory. Since the coefficients are in general complex, it means that the contribution of any pole to the response of the system can be determined by means of its position on the complex $s$-plane. This is illustrated in figure 2.3. The most important observation to be made from the figure is that any poles in the right hand half plane represent an unstable system, since there is a growing form of disturbance, either sinusoidal or exponential, independently of any system input. Thus a major constraint on any usable system, other than an oscillator, is that it should have no poles in the right hand half plane.

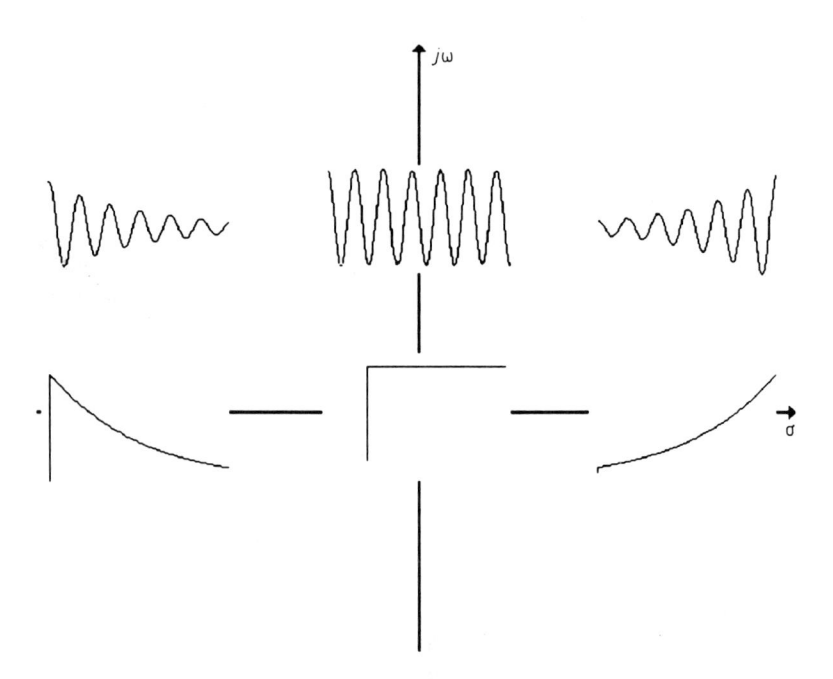

**Figure 2.3** Illustration of the response induced by poles in different parts of the $s$-plane.

To a great extent systems may be analysed and synthetized by means of the manipulation of poles and zeros. An important example is the design of filters, one at least of which will be found in the intelligent sensor. Many primary sensors will reveal a single pole (damped) response, and indeed this behaviour was often designed in to provide a crude means of combating noise. Other sensor mechanisms (e.g. accelerometers) are second order systems which will reveal an oscillatory behaviour from a pair of complex poles.

### 2.1.7 Convolution and deconvolution

One of the most important ways in which sub-systems are assembled into systems is by means of cascading, and as we shall see in the next chapter sensors are no exception. For example, a gramophone (phonograph) comprises a set of sub-systems: the pick-up, which is the input transducer, the amplifier, and the loudspeaker, which is the output transducer. These are all connected in cascade. In the time domain the cascaded connection gives rise to a rather complex relationship. Indeed, this relationship also applies to the determination of the output of the system to an arbitrary input signal. This is the convolution integral. Thus

$$y(t) = h(t) * x(t) = \int_0^\infty x(t-\tau) h(\tau) \, d\tau = \int_{-\infty}^t x(\tau) h(t-\tau) \, d\tau \, . \quad (2.21)$$

One of the major reasons for the importance of the frequency domain is the fact that this difficult operation under the Laplace or Fourier transform becomes a simple one of multiplication, i.e. we have the convolution theorem

$$Y(s) = L[\, h(t) * x(t) \,] = H(s) \, X(s) \, . \quad (2.22)$$

So, for example, to find the overall response of our gramophone, in the time domain we would need to perform a complex piece of integration, while in the frequency domain we simply multiply together the frequency responses of the separate components. In fact the representation is even simpler with poles and zeros. For, in this case, convolution is simply replaced by the superposition of the pole-zero diagram of each of the components of the system to form an overall pole-zero diagram.

The determination of the frequency response of a system from the $s$-domain representation, or indeed the pole-zero representation, is simply a matter of substituting $s = j\omega$. In geometrical terms, we can imagine this process in the form of a representation of the system response as an elastic surface above the $s$-domain. Poles may be thought of as pushing this surface up, and zeros as pinning it down, The frequency response as a graph of amplitude against frequency is then the section of this surface as seen along the $j\omega$ axis. Thus, for example, if a pair of complex poles approach the $j\omega$ axis, there is a peak in the response, which is the phenomenon of resonance. Many important sensor mechanisms exhibit this phenomenon, and dealing with it is a major function of sensor signal processing.

While convolution is fundamental to the operation of systems, its inverse, deconvolution, is often a very important concept in the utilization of practical systems, especially of transducers. Deconvolution is the answer to the question 'given the output of the system, what was the input?'. The operation of deconvolution, or equalization, is very difficult to realize in traditional linear continuous systems. In the time domain, it is represented by the difficult solution of the integral equation. In the $s$-domain, it is realized by a system which is the reciprocal of the given system. This is not necessarily easy or even possible to realize (indeed, one of the great limitations of classical linear continuous systems is this realization problem; we may have a mathematical model for a desired system, but it does not necessarily follow that we can produce a practical piece of hardware that will conform to that model or will not drift away from the optimum). On the pole-zero diagram the deconvolution process is represented by the superposition of a pole on every zero and a zero on every pole. Again this is very difficult in general to realize in terms of classical linear continuous systems, but it can be relatively easy, as we shall see, in discrete or digital systems.

### 2.1.8 Signals as stochastic processes

The theory of randomness in signals and systems (Papoulis 1965) is very important in applied electronic engineering, since:
1) A purely determined signal contains no information whatsoever, so all informative signals must, to some extent, be considered as random events.
2) All practical systems are affected by noise, if only because they share the situation of the rest of the universe by being in a state of continual thermal agitation.

A stochastic process may be thought of as a family of functions of time, one of which happens to be the particular outcome we are dealing with as our signal, and the members of this family (which is usually infinitely large) are characterized by certain common overall average properties which are conditioned by their origin and subsequent treatment. These properties, which we can only estimate as we do not have access to the whole family, thus convey information about the signal source, and the retrieval of such information is often one of the objectives of instrumentation engineering. Often we require to extract some number or function which epitomizes some aspect which is important to us. Examples of such numbers are RMS amplitude, bandwidth, dominant periodicity, etc. Examples of functions are those important ones which condense the properties of the signal with respect to amplitude, time and frequency. They are respectively the distribution, the autocorrelation and the power spectrum.

## SIGNALS AND SYSTEMS

The distribution of the signal is a means of condensing into a single function those properties of a signal which are independent of the time scale. It is defined by

$$F(x) = \text{Prob}\,[\,x(t) \leq x\,] \tag{2.23}$$

where $x(t)$ is the instantaneous signal amplitude and $x$ is a number, the argument of the function $F(x)$. In non-probabilistic terms it can be thought of as the fraction of time for which $x(t) < x$.

More commonly used is the density $f(x)$, which may be derived from $F(x)$ by

$$f(x) = \frac{d}{dx} F(x) \,. \tag{2.24}$$

An important, indeed vital, application of the density function; in intelligent sensor theory is in the selection of gain prior to AD conversion, since it determines how much of the signal will be lost by limiting.

A number of important functions may be represented in terms of averages, either the ensemble average of a function $g(x)$ of $x$

$$E\{g(x)\} = \int_{-\infty}^{+\infty} g(x)\, dF(x) \tag{2.25}$$

or the time average

$$\overline{g(x)} = \lim_{T \to \infty} \left\{ \frac{1}{2T} \int_{-T}^{+T} g\,[x(t)]\, dt \right\} \,. \tag{2.26}$$

When $E\{g(x)\} = \overline{g(x)}$ the system is said to be **ergodic**.

Some definitions of important averages are:

the mean value $E\{x\}$,

the mean square value $E\{x^2\}$,

the autocorrelation $R_{xx}(t_1, t_2) = E\{x(t_1), x(t_2)\}$ .

This last definition also allows us to give one definition of the term **stationarity** in the wide sense, by saying that a process is stationary if the autocorrelation can be expressed as a function of one variable, the time difference, i.e.

$$R_{xx}(t_1, t_2) = R_{xx}(\tau) \quad \text{where } \tau = t_1 - t_2 \,. \tag{2.27}$$

Non-stationarity is not an easy concept to deal with, but in instrumentation we are often obliged to grapple with it. One way intelligent sensors allow us to do this is by making gain and offset adaptive functions.

If we have a second stationary process $y(t)$ we can form a function which represents the statistical relationship between the two signals, the cross-correlation

$$R_{xy}(\tau) = E\{x(t+\tau)\,y(t)\} = E\{x(t)\,y(t-\tau)\}\,. \tag{2.28}$$

An important result if $x$ and $y$ are respectively the input and output of a linear system of impulse response $h(t)$ is that

$$R_{xy}(\tau) = R_{xx}(\tau) * h(t) \tag{2.29}$$

and if $x$ is uncorrelated (noise-like) the cross-correlation is a reproduction of the impulse response.

Another important function describing the properties of a signal is the power spectrum. This is only properly defined in terms of the Fourier transform of the autocorrelation.

$$S_{xx}(\omega) = F\{R_{xx}(\tau)\}\,. \tag{2.30}$$

Other definitions are often applied by engineers, but they do not hold up mathematically. The power spectrum is important in many sensor applications (e.g. laser Doppler velocimetry). These functions are all important in sensor applications, and the measurement process is often one of encapsulating some property of the target physical signal in one or more of these forms.

As an example let us consider a triangular waveform as shown in figure 2.4 (a). First of all we evaluate the distribution $F(x)$. In order to do this we only have to look at one triangle, as the same calculation applies to all the rest. Remembering that $F(x)$ is defined as the probability that $x(t) < x$, we can determine that for a triangle of duration $T$ and peak amplitude $A$, $F(x)$ is the proportion of time for which $x(t) < x$. By similar triangles we deduce that $F(x) = x/A$ for $0 \leq x \leq A$. Further we can deduce that the density $f(x) = dF/dx = 1/A$, i.e. we have a uniform distribution. Furthermore these calculations apply whatever the duration of the triangle, $T$, so that a waveform consisting of triangles of constant amplitude, $A$, but random duration as in figure 2.4 (b) will exhibit exactly the same distribution and density. This happens to be an example that will be useful later.

Let us calculate some averages in the two possible ways; first the simple mean. The ensemble average is given by

$$E(x) = \int_0^A x\,dF(x) = \int_0^A x\,\frac{dF}{dx}\,dx = \int_0^A x f(x)\,dx = \int_0^A \frac{x}{A}\,dx = \frac{A}{2}\,. \tag{2.31}$$

# SIGNALS AND SYSTEMS

The time average is given by

$$\overline{x(t)} = \frac{1}{T}\int_0^T x(t)\,dt = \frac{1}{T}\int_0^T \frac{tA}{T}\,dt = \frac{A}{2}. \qquad (2.32)$$

The fact that these two estimates are equal indicates that the process is ergodic. Similarly we can deduce for the second order statistic that

$$E(x^2) = \overline{x^2} = \frac{A^2}{3}. \qquad (2.33)$$

Thus the triangular waveform is ergodic even if the length of the triangles varies at random or systematically, as in figures 2.4 (b) and 2.4 (c).

Considering the waveform as an oscillatory function without the mean (or zero frequency component) we have the mean square value

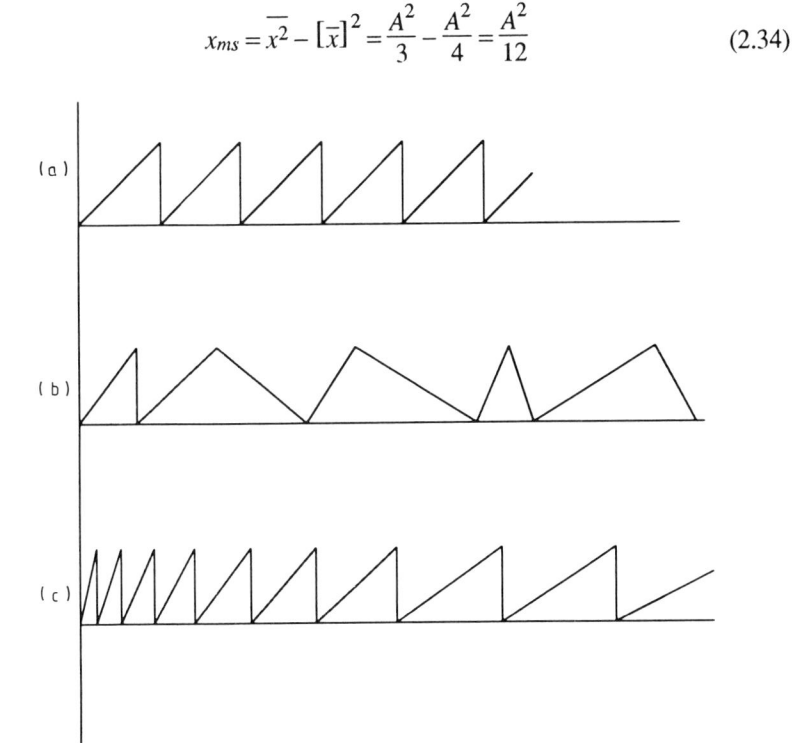

$$x_{ms} = \overline{x^2} - [\overline{x}]^2 = \frac{A^2}{3} - \frac{A^2}{4} = \frac{A^2}{12} \qquad (2.34)$$

**Figure 2.4** Three examples of waveforms composed of triangles.

and hence the RMS value

$$x_{\text{rms}} = \frac{A}{2\sqrt{3}}.  \quad (2.35)$$

It is interesting to note that these calculations apply even in a situation like that shown in figure 2.4 (c), where the process is evolutionary. This is, in fact, a process that is ergodic while not being stationary.

The autocorrelation is the function that summarizes the properties of the signal as they are distributed in time. In figure 2.4 (a), in the range $0 \leq x \leq T$, we can evaluate from the definition as a time average

$$R_{xx} = A^2 \left\{ \frac{\tau^2}{2T^2} - \frac{\tau}{2T} + \frac{1}{3} \right\}  \quad (2.36)$$

and we note that it is periodic with period $T$. This is, indeed, the basis of an important class of intelligent sensing processes in which it is necessary to identify a periodic signal immersed in an aperiodic background. We also note that by definition $R_{xx}(0) = \overline{x^2} = A^2/3$. The equal maxima of $R_{xx}$ occur at multiples of $T$, and the minima of the parabolae lie half way between them.

We can create a number of useful examples from the triangular case. In the case of figure 2.4 (b), where the lengths of the triangles are random, we can only calculate $R_{xx}$ if there is information about the distribution of lengths. We have seen the case of figure 2.4 (c), where the triangles are of constant amplitude but slowly increasing length $T$, which gives us a system that can be ergodic but not stationary. The triangular waveform will also be important in our discussion of quantization (§ 4.5.2).

## 2.2 SAMPLED SIGNALS AND DISCRETE SYSTEMS

### 2.2.0 Introduction

Traditional linear continuous electronic systems operate by means of the continuous redistribution of voltages and currents in electronic networks. Digital systems, in contrast, operate by means of discrete changes in the states of bistable sub-systems. The substitution of a two-dimensionally discrete representation for the former continuous representation in both amplitude and time is a very fundamental change, the consequences of which must not be underestimated (Mayhan 1983, Oppenheim et al 1983). In some applications of sensors the input information is already in discrete or number form. While the following theory does not apply directly to such applications, it may be important indirectly, an example being the effect some sequence of arithmetic operations may have on a set of numbers in general; in this respect such aspects as instability may be significant. For applications where the discrete processor is dealing essentially with continuous real time information, however, the sampling theory is absolutely vital. It is essential for any engineer dealing with such applications to appreciate the constraints of the theory. In this section we shall be dealing with one dimension of the discreteness, that is the time dimension.

### 2.2.1 The sampling process

In the present context the processor operates in a sequential manner, and so, if it is to deal with continuous information, it is necessary to freeze such information temporarily while the device goes through its prescribed sequence before another datum can be accepted. This process of representing an infinite set of points forming a time segment of a signal by one point is a profound distortion, and it involves throwing away information which can never be retrieved again. Mathematically, this process of sampling can be conveniently represented by the sampling property of the delta function. Thus, in the case of a signal $f(t)$ being sampled as a time $t_1$, we have

$$f(t_1) = \int_{-\infty}^{+\infty} f(t)\, \delta(t-t_1)\, dt \; . \tag{2.37}$$

In the majority of signal processing applications it is desirable to sample the signal at regular intervals. This is achieved mathematically by extending the single delta function into a train of delta functions, the so-called *comb* function, which we represent by the Greek symbol $\Xi$. Thus

$$\Xi(t) = \sum_{n=-\infty}^{+\infty} \delta(t-n) \; . \tag{2.38}$$

Now these special functions $\delta(t)$ and $\Xi(t)$ only have meaning within an integral, where they come into operation by means of their sampling properties. We have seen that in terms of signal theory advantages may be obtained by transformation into the complex frequency domain. So let us examine the Laplace transform of such a sampled signal. This may be written

$$F(s) = \int_{-\infty}^{+\infty} \Xi\left(\frac{t}{T}\right) f(t)\, e^{-st}\, dt \tag{2.39}$$

where $T$ is the sampling interval. Thus the complex frequency domain representation is simply, by application of the sampling property and the Laplace shift theorem, equation (2.42)

$$F(s) = \sum_{-\infty}^{+\infty} f(nT) \, e^{-snT} \, . \tag{2.40}$$

This relationship as we shall see leads to a powerful mathematical tool for the analysis and synthesis of digital systems.

### 2.2.2 Hold

Before we go on to examine the further mathematical implications of sampling, it is important to underline the implications of discarding the information between samples. This can never again be retrieved save in certain situations where the properties of the signal are constrained. Therefore in order to return the signal to the outside world, even if we have performed no process upon it, we are forced to make some assumption about what has gone on in between sampling points.

This is the electronic process of digital-to-analogue conversion. Nearly always, for the sake of electronic simplicity, we make the simplest possible assumption; that is, that the signal retains the level dictated by the last sample received until the next sample is received. This is known as zero order hold. It must be emphasized that the process of sampling followed by zero order hold is a gross distortion of the signal, and we will examine this later in the discussion of quantization.

### 2.2.3 The z-transform

This transform, which is absolutely fundamental to the analysis and synthesis of real time signal processes, is nothing more than the sampled form of the Laplace transform simplified by the substitution $z = e^{sT}$.

Thus we see that from equation (2.40), any sampled signal can be represented in polynomial form by

$$F(z) = \sum_{n=-\infty}^{+\infty} f(nT) \, z^{-n} \, . \tag{2.41}$$

Equally the transform of the response of a causal system can be written

$$H(z) = \sum_{n=0}^{\infty} h(nT) \, z^{-n} \, . \tag{2.42}$$

So all signals and linear systems can be represented by polynomials in $z$ and the transformation is simply achieved by setting the coefficient of $z^{-r}$ equal to the $r^{th}$ sample ordinate of the signal or response function.

Hence we have created a third domain, the $z$-domain, which is suitable for manipulation of sampled systems and signals. A number of simplifications arise from the use of this transform. For example, one feature is the form taken by the shift theorem. In Laplace transform terms, a shift theorem, which arises directly from the definition of the transform tells us that if $y(t) = x(t - rT)$ then

$$Y(s) = e^{-srT} X(s) \qquad (2.43)$$

and in the $z$-domain this becomes simply

$$Y(z) = z^{-r} X(z) \qquad (2.44)$$

so the operation 'delay $r$ sample-times' is represented by multiplication by $z^{-r}$. Hence each term of the polynomial in the defining equations above represents a combined delay and weighting operation, which is consistent with the role played by the function $h$ in convolution with a signal. The operation of deconvolution is rather inaccessible in the time domain since it demands the solution of an integral equation, but it becomes relatively simple in the $z$-domain. We can tabulate the mappings of important operations between the three domains.

| $t$ | $\omega$ | $z$ |
| --- | --- | --- |
| addition | addition | addition |
| multiplication by a constant | multiplication by a constant | multiplication by a constant |
| Convolution | Function multiplication | Polynomial multiplication |
| Deconvolution | Function division | Polynomial division |

The first two lines of the table tell us that the mappings are linear. The last two lines show us that the unravelling as well as the ravelling of signals with system functions is feasible in digital form.

The mapping between the $z$ and $s$-domain is embodied in the defining substitution $z = e^{sT}$, and it is rather important to understand its implications, which are summarized in figure 2.5. The important features to note from this figure are:

- the $j\omega$ axis is mapped into the Unit circle,
- w is naturally expressed as a fraction of the sampling frequency $\omega_o = 2\pi/T$,

- the left-hand half plane (of stable poles) is mapped to the inside of the unit circle,
- the mapping from $s$ to $z$ is many to one,
- The mapping is unique only if all $\omega \leq \omega_s/2$,
- any $\omega > \omega_s/2$ has an indistinguishable alias in the range $0 \leq \omega \leq \omega_s/2$,
- $e^{\sigma T}$ and $\omega T$ become respectively the magnitude and angle of $z$.

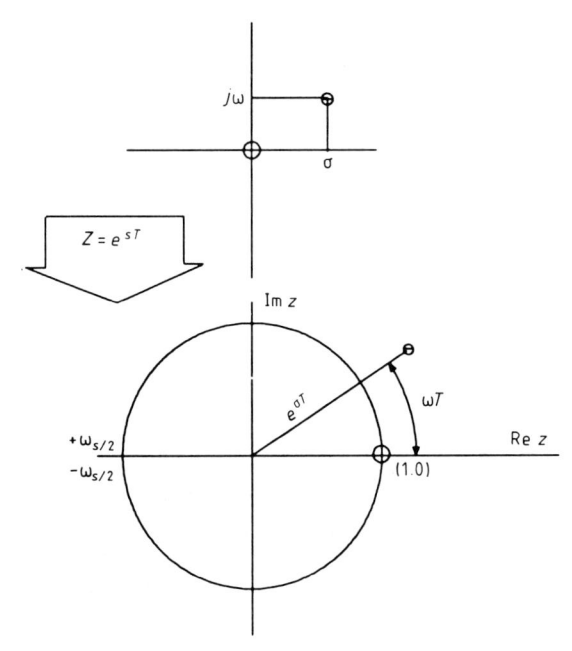

**Figure 2.5** Geometrical interpretation of the $z$-transform.

## 2.2.4 Sampling Theorem

We have observed above that it is possible for a signal to have an alias once it is sampled. That is to say, sinusoidal signals of different frequencies may produce an identical sampled signal. This point is so important as a restriction on the real-time processing of continuous signals by discrete processors that it is worthwhile to examine it from another angle.

Let us consider two sinusoidal signals, and for simplicity we will neglect the problem of phase. If the frequencies of the signals are $\Omega_1$ and $\Omega_2$, and they are sampled at times $nT$, (where $n = .... -2, -1, 0, +1, +2, +...$) then the signals are aliases if they are equal at all sample points; i.e.

$$\cos(n\Omega_1 T) = \cos(n\Omega_2 T) . \qquad (2.45)$$

By a well known trigonometrical relationship we can convert this to:

$$\sin\left\{n(\Omega_1 + \Omega_2)\frac{T}{2}\right\} \sin\left\{n(\Omega_1 - \Omega_2)\frac{T}{2}\right\} = 0 \text{ for all } n. \qquad (2.46)$$

This is true if $\Omega_1 \pm \Omega_2 = 2\pi k / T$, where $k = 0, 1, 2, 3,...$ When $k = 0$ the frequencies are equal and the signals are identical. When $k = 1$ we have the first alias, i.e. the lowest frequency for which the samples are identical. This has the simple representation in the z-domain of the reflection in the real axis. Higher values of $k$ produce either the same point on the unit circle as the original frequency or the reflection point, according to the sign in the above equation.

The sampling theorem tells us that, if ambiguity is to be avoided, either the sampling frequency must be set at a value at least twice the highest frequency present, or the signal must be filtered so that this proscription is obeyed. It is important to note that the sampling theorem applies not only to a signal we may be interested in, but it also applies to any noise present. In certain circumstances the aliasing of high frequency noise into the range of observation can cause serious errors, so the pre-filtering of signals for sampling would normally be considered an essential requirement of a the digital processing sub-system of an intelligent sensor. As a result the bandwidth of a digital system is very much controlled by the amount of processing which can be carried out between samples (if only on average).

### 2.2.5 A simple process

As a simple example let us consider a process, which is (perhaps misleadingly) easy to implement in a microprocessor, smoothing by the method of the running mean (or moving average).

In the situation where a low frequency signal is apparently contaminated by random variations of a high frequency nature, it is not uncommon to find that a programmer will spontaneously produce a method such as the running mean, i.e.

$$y_k = \frac{1}{n} \sum_{r=0}^{n-1} x_{k-r} . \qquad (2.47)$$

Until the filtering interpretation was understood, $n$ was apparently chosen by a combination of faith, hope, trial and error. The method appeared to be successful, perhaps more so with some form of periodic disturbance than others, but as intuition would suggest it certainly reduced short-term fluctuations of data, and one never seemed to find time to question the exact implications of the procedure. The development of techniques such as the $z$-transform has made this fairly easy, for the iteration above is obviously a non-recursive filter of the form

$$H(z) = \frac{1}{n} \sum_{r=0}^{n-1} z^{-r} \quad . \tag{2.48}$$

The important conceptual advance is the understanding that the operation is independent of $k$. This is a good example to illustrate the greater efficiency often obtained by changing to a recursive scheme, since we may re-write it as a geometric progression

$$H(z) = \frac{1}{n} \frac{1-z^{-n}}{1-z^{-1}} \quad . \tag{2.49}$$

The inverse transform is

$$y_k = y_{k-1} + \frac{1}{n}(x_k - x_{k-n}) \tag{2.50}$$

which obviously yields the same result as the first equation above. We are led to wonder how much computer time has been wasted in the past by the computation of the redundant central terms in this first equation.

Such profligacy is certainly not tolerable in real-time work. How does the operation treat the different frequency components of the signal? We can find out by examining it on the unit circle, by substituting $z = e^{j\omega T}$ whence the recursive scheme yields

$$H(\omega) = \exp\left[-(n-1)\frac{j\omega T}{2}\right] \cdot \frac{\sin(n\omega T/2)}{n \sin(\omega t/2)} \quad . \tag{2.51}$$

Figure 2.6 shows that the response of this formula has peaks and zeros, and it is in fact in the form of a *sinc* function. Thus the apparently simple scheme smoothing by the running mean is in fact something with a rather complicated frequency response, and its success with a given periodic disturbance may well depend upon the choice of $n$ with respect to the frequency of that disturbance.

In fact the above result is a manifestation of the most important Fourier transform function pair, the block function and the *sinc* function (figure 6.14). One of the consequences of this pair is that if we take a finite block of samples to represent

# SIGNALS AND SYSTEMS

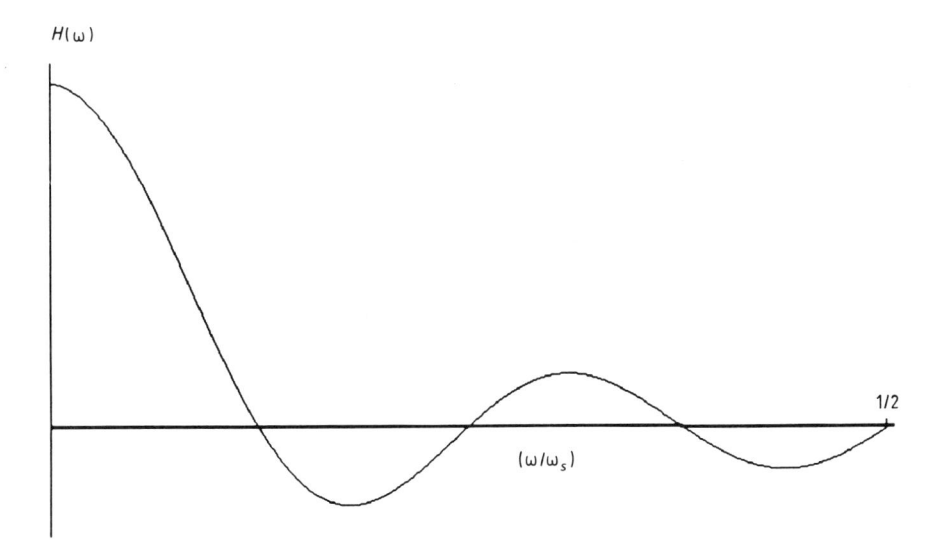

$H(\omega)$

$(\omega/\omega_s)$

1/2

**Figure 2.6** Frequency response of the running mean process for $n=8$.

the whole of an infinite time series the effect is to convolve the frequency response with a *sinc* function, which introduces irrelevant frequency lobes into the spectrum. This introduces the so-called window problem and means that there can be no unique estimate of the spectrum. Indeed, in a way that is analogous to the uncertainty principle in physics, the more we locate a function in the time domain, the less we locate it in the frequency domain and vice versa, as we shall see later in § 6.3.4.4.

It is interesting to note that since one of the authors used this simplest of examples in a text book in 1975 not a year has gone by without his finding a graduate student or industrial engineer using the running mean with calculation of the redundant central terms and not realizing that it has such a marked frequency response function.

### 2.2.6 First and second order systems

Our simple model of poles and zeros in terms of a surface which is pushed up or pinned down tells us that system responses are dominated by those roots that are closest to the imaginary axis. In many ways the second order system is the natural pattern of things. All electrical systems must have inductance, capacitance and resistance and all mechanical systems must have inertia, springiness and damping. In traditional instruments the problems imposed by such responses were overcome by overwhelming them with a dominant real pole. A simple moving coil galvanometer in its basic form was virtually unusable because in any practical application the pointer would oscillate so uncontrollably that it was unreadable. The solution was to provide damping by, for example, a dashpot. This had the effect of imposing a pole that was so dominant that all other responses were negligible. A one pole system corresponds to the first order differential equation:

$$y(t) + \tau \frac{dy}{dt} = K x(t) \tag{2.52}$$

which in the $s$-domain gives us

$$Y(s) = \frac{K}{1 + s\tau} X(s) \tag{2.53}$$

while in the $z$-domain, by matching poles and zeros we have an equivalent system

$$Y(z) = \frac{z}{z - z_1} X(z) \tag{2.54}$$

where $z_1 = exp(-T/\tau)$, and inverting we get a discrete equation which behaves in the same way

$$y_i = x_i + z_1 y_{i-1} \,. \tag{2.55}$$

Nevertheless the most pervasive form of response in sensors, and particularly mechanical ones, is the second order response. This yields a differential equation of the general form

$$a \frac{d^2 x}{dt} + b \frac{dx}{dt} + c = y(t) \tag{2.56}$$

the equivalent polynomial equation in the $s$-domain being

$$(a s^2 + b s + c) X(s) = Y(s) \tag{2.57}$$

while the difference equation which behaves in the same way as the continuous system is

$$A\, x_{i-2} + B\, x_{i-1} + C x_i = y_i \ . \tag{2.58}$$

The relationship between the various coefficients follows from the equivalences established above, and will be pursued later in some of the more practical examples. The powerful aspect of the introduction of the $z$-transform is that, as we see from the table in § 2.2.3, the inverse system is obtained simply by taking the reciprocal of the $z$-transform, which means that we can create a digital system which completely negates the undesired responses from the continuous system, which without intelligent sub-systems would be an immutable characteristic of the sensor.

### 2.2.7 Root loci

An important concept which we shall need in the treatment of sensor compensation is the root locus, i.e. the path taken by a pole or zero as a parameter of the system changes. Consider the simple electrical system of a series $RLC$ circuit. By treating the inductor and capacitor as resistances of value $sL$ and $1/sC$ respectively we can write the $s$-domain equation by inspection:

$$\left( sL + \frac{1}{sC} + R \right) I = V \tag{2.59}$$

and the admittance of the circuit is

$$Y = \frac{I}{V} = \frac{s/L}{s^2 + \frac{R}{L} s + \frac{1}{LC}} \ . \tag{2.60}$$

As the system is second order there are two poles. Now as an example of loci consider the behaviour of these roots as $R$ changes from 0 to $\infty$. when $R = 0$ the poles are at $s = \pm\, 1/\sqrt{LC}$. As $R$ increases they move round a semicircle of radius $1/\sqrt{LC}$ until they meet at the point $s = -\, R/2L$. Then as $R$ increases towards infinity the poles separate and move in opposite directions one towards zero and one towards $-\infty$. We can see from our tables of Laplace transforms that when $R = 0$ the response of the system is an undamped oscillation. As it increases the oscillation becomes more and more damped until, at the critical point $s = -\, R/2L$, it is critically damped with an impulse response of the form $t\, e^{-t}$, equation (2.10). Thereafter it is overdamped with the response being the sum of two decaying exponentials.

Resonant systems can also be represented in general terms by the meaningful quantities of a characteristic frequency $\omega_o$ and the 'quality' factor, $Q$. In which case the poles are at

$$s = -\frac{\omega_o}{2Q} \pm j\omega_o \left(1 - \frac{1}{4Q^2}\right)^{1/2}. \qquad (2.61)$$

For the series tuned circuit, $Q = \omega_o L/R$ and $\omega_o = 1/\sqrt{LC}$. If $\omega_o$ is constant, $Q$ may be considered the parameter which defines the loci in figure 2.7, and the critical point occurs when $Q = 1/2$. There are, in fact, three possible definitions of resonant frequency (maximum response, zero phase and natural frequencies) which may or may not be equal, but in high $Q$ systems they converge and the poles are found at

$$s = -\frac{\omega_o}{2Q} \pm j\omega_o. \qquad (2.62)$$

As we shall see, resonance is an important phenomenon, for both positive and negative reasons, and the concept of the root locus is an important tool in dealing with it.

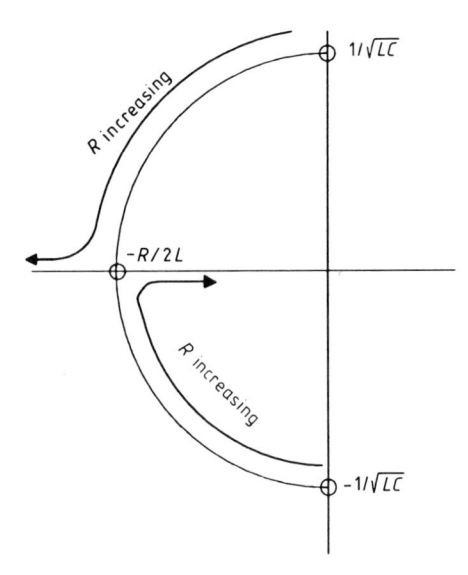

**Figure 2.7** Locus of the poles of a series tuned circuit as the resistance varies.

# 3
# Physical Principles of Sensing

## 3.1 GENERAL PRINCIPLES OF TRANSDUCTION

We have seen that the process of measurement is one of assigning numbers to events and entities in the real world. That world comprises a dynamic arrangement of different forms of energy. Of these, matter, according to relativity theory, may be considered a particular manifestation. Matter, however, has an important property which is the sole basis of transduction: it is able to convert changes in one form of energy into changes in another form. A trivial example would be the conversion of gravitational potential energy into mechanical kinetic energy. Primary transducing elements, of which primary sensing elements are an important sub-set, thus comprise lumps of matter. The type of primary transducer they represent is controlled by the form of the lumps and/or the materials which go to make them up. This may be illustrated by considering a random selection of examples:
    Windmill
    Thermocouple
    Oar
    Tinder box
    Solar cell
Evidently sometimes the shape of the lump is most important and sometimes it is the composition: often it is both.

Unfortunately, there is some variation of usage in the use of terms such as transducer, sensor, detector and actuator. Some authorities contend that the term transducer should only be applied to energy conversion devices, and that sensors are something different. In this treatment we shall take the view that such a distinction is trivial, in that we are concerned with the conveyance of information, and information cannot be conveyed without the passage of energy, however small the amount. The problem does not arise with an actuator, which by its very nature is concerned with producing energetic effects in the outside world.

Thus we define our terms with reference to our measurement or control system. **Transducers** divide into two sub-sets, **sensors** which input information into our system from the external world and **actuators** which output actions into the external world. **Detectors** will be regarded as binary sensors. This usage is simple, unambiguous and adequate for our purposes.

### 3.1.1 Physical variables

When we examine any one particular form of energy we find that it is represented by a pair of variables - a potential (or across) variable and a flux (or through) variable (Finkelstein and Watts 1971). The familiar examples are voltage and current. Which is which has nothing to do with cause and effect, but is determined by which of two continuity laws they obey. Potential variables obey a law of equilibrium round a path (e.g. Kirchhoff's Voltage law) while flux variables obey a law of equilibrium at a point (e.g. Kirchhoff's Current Law).

Table 3.1 shows some examples of corresponding potential and flux variables in various energy systems. The analogy with voltage and current is an important one, as it give us a powerful model for the primary transduction element in the form of the familiar two-port system as shown in figure 3.1.

Table 3.1. Classification of basic physical variables

|  | Flux ('through' variable) | | Potential ('across' variable) | |
|---|---|---|---|---|
|  | State | Rate | Rate | State |
| General (basic relationship) | $y$ | $\dot{y} = \dfrac{dy}{dt}$ | $\dot{x} = \dfrac{dx}{dt}$ | $x$ |
| Mechanical-translational | Momentum | Force | Velocity | Displacement |
| Mechanical-rotational | Angular momentum | Torque | Angular velocity | Displacement |
| Electrical | Charge | Current | Voltage | Flux linkages |
| Fluid flow | Volume | Flow rate | Pressure | - |
| Thermal | Heat | Heat flow rate | Temperature | - |

The characteristics of the electrical two-port are that the nature of any source or load connected to one port in general affects the behaviour at the other port. The theory of two-ports is familiar to electronic engineers and was highlighted when the transistor began to replace the vacuum tube, whose behaviour was somewhat simpler. The essential problem facing engineers at that time was which set of parameters to adopt in order best to describe the behaviour of transistors. The nub of this problem is the fact that four variables (and the four parameters which

*PHYSICAL PRINCIPLES OF SENSING* 35

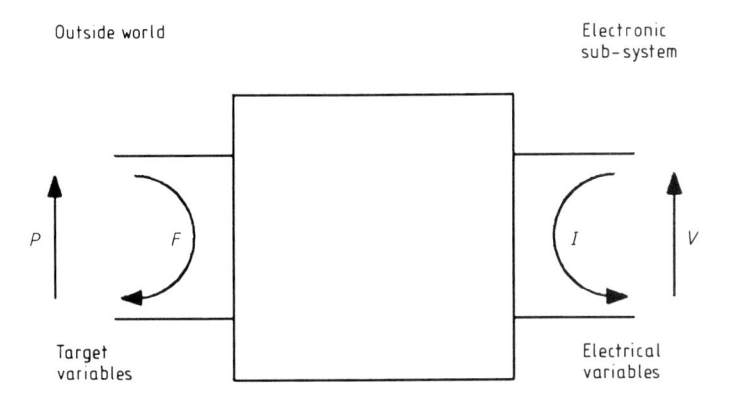

**Figure 3.1** The sensor as a two-port network.

inter-relate them) can be arranged in twelve different independent permutations, each of which represents a possible parameter system.

With transistors the hybrid parameter system, which initially seemed somewhat bizarre, was widely used as the parameters had relevance to a physical model of the device, but for the theory of transducers a more appropriate form is the cascade parameter system. The advantage of this system is that the cascading of two or more subsystems is represented by the multiplication of their corresponding matrices of parameters. Note that in making this statement we are making a presumption of linearity, without which linear algebra including matrices breaks down, and in general the non-linear case is much more complicated. These ideas are illustrated in figure 3.2.

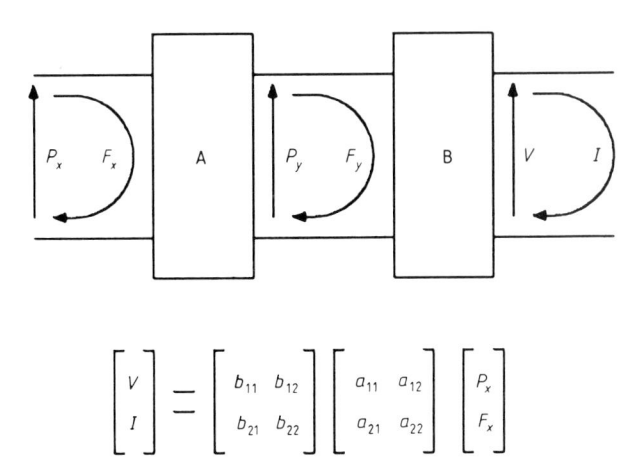

$$\begin{bmatrix} V \\ I \end{bmatrix} = \begin{bmatrix} b_{11} & b_{12} \\ b_{21} & b_{22} \end{bmatrix} \begin{bmatrix} a_{11} & a_{12} \\ a_{21} & a_{22} \end{bmatrix} \begin{bmatrix} P_x \\ F_x \end{bmatrix}$$

**Figure 3.2** The primary sensor as cascaded sub-systems.

It should be observed that even the simplest of devices is in effect a set of cascaded sub-systems. A good example is a simple piece of wire, which when heated by an electric current becomes a hot-wire anemometer (see § 3.3.1.5). Thus flow variables (pressure and mass flow) are translated to thermal variables (temperature and heat flow) and thence to electrical variables (voltage and current).

In the ideal world the system parameters would all be linear and time-invariant, and fortunately we can often make this approximation, but it should only be made with careful consideration. In fact our example of the simplest of structures, the hot wire anemometer, is a non-linear transducer, and special methods are needed to deal with this characteristic. A more universal problem is that the parameters are invariably frequency dependent (the time domain representation being virtually useless in this case, as it would imply multiple convolutions). In specific applications it may be that over the bandwidth of the target signal these variations are negligible, but again it should never be assumed that this is so without evidence. Defects in the frequency response and their treatment are important topics in the theory of transducers, and the need to cope with them in the light of the limitations of classical electronic techniques represent one of the more powerful motivations towards the intelligent transducer approach (see § 6.4.3).

## 3.2 PRIMARY SENSOR DEFECTS AND THEIR COMPENSATION

We have already referred to some defects found in primary sensor mechanisms, and it is useful to list the five major ones (Brignell 1987a). They are
    1) Time (or frequency) response
    2) Non-linearity
    3) Noise
    4) Parameter drift
    5) Cross sensitivity.

As we have observed, **time (or frequency) response** is a universal problem. The reason for this is the co-existence of dissipative, storage and inertial elements (the analogues of resistance, capacitance and inductance) in the lump of material that is our primary sensor. These translate into the time derivatives appearing in the differential equation that models that system and hence powers of $s$ in the corresponding ratio of $s$ polynomials. In principle frequency correction is a simple matter. If our device has a complex frequency response $H(s)$ then we simply have to cascade this with a filter of response $G(s)$ such that $H(s).G(s) = 1$ (or more practically $=exp[-sT]$ where $T$ is the propagation delay).

In pole-zero terms this means we place a pole over every zero and a zero over every pole so that they all mutually cancel. In practice the realization of such a filter by classical means will often represent a non-trivial task, and there are related problems such as stability and component sensitivity which complicate matters.

# PHYSICAL PRINCIPLES OF SENSING

The realization of digital filters with given poles and zeros is relatively straightforward; though it does not follow that the processing power at hand is sufficient to implement it for the bandwidth required.

Another feature of digital filters is that they require a constant sampling frequency to be maintained. This can complicate matters when other slower processes (such as monitored temperature compensation) have to coexist with them. In this case we have to provide a mechanism for the slower process to be interleaved with the regularly sampled process so that it is carried out piece by piece over a number of samples. The simplest means of achieving this is by having the sampling and digital filtering operating as an interrupt program, with the slower processes being carried out as the main program loop. Unfortunately the overheads of computation required to implement this switching from one task to another add to the overall load, and as a result the available bandwidth can be diminished by a greater amount than would be implied by simple consideration of the individual tasks.

It is implied in the above that in order to produce a response-correcting digital filter we have to have a model of the individual sensor (in, say, the form of a pole-zero diagram). This may be obtained by theoretical analysis or, much preferably, by measurement. However, all models being imperfect, this is still a hazardous situation, and when we come to discuss the implementation of such filters utilizing the available intelligence the search for model-free methods will be an important issue (§ 6.4.3).

It is difficult to remember what a dominating defect **non-linearity** was in the days of linear continuous electronics. There were some techniques available (such as the use of diode networks to realize reciprocal characteristics), but these were relatively crude. As a result non-linear primary sensors tended to be ignored except in particular circumstances (e.g. where the sensor is used in a nulling mode at one particular operating point). With digital electronics the problem is greatly diminished, and linearization processes such as Look Up Tables (LUTs) or polynomials are now well established (§ 6.2.2.1).

Nevertheless it is important to note that some characteristics are inherently pathological in that no amount of processing will correct them without great loss of accuracy. Typical of such characteristics are ones which show a 'dead-band', i.e. a region of small or zero slope. This is illustrated in figure 3.3 which shows how a region of low slope in the input/output characteristic of the sensor causes the quantization levels of the output voltage to be mapped back to unacceptable inaccuracies in the measured variable (and the same consideration applies to other forms of voltage noise).

We can express the phenomenon in figure 3.3 algebraically in terms of the characteristic of the sensor, say $v = f(x)$. If there is an error in the voltage, $e_v$, be it quantization or any other form of noise, then the error in the target variable is given by

$$e_x = \frac{e_v}{f'(x)} \tag{3.1}$$

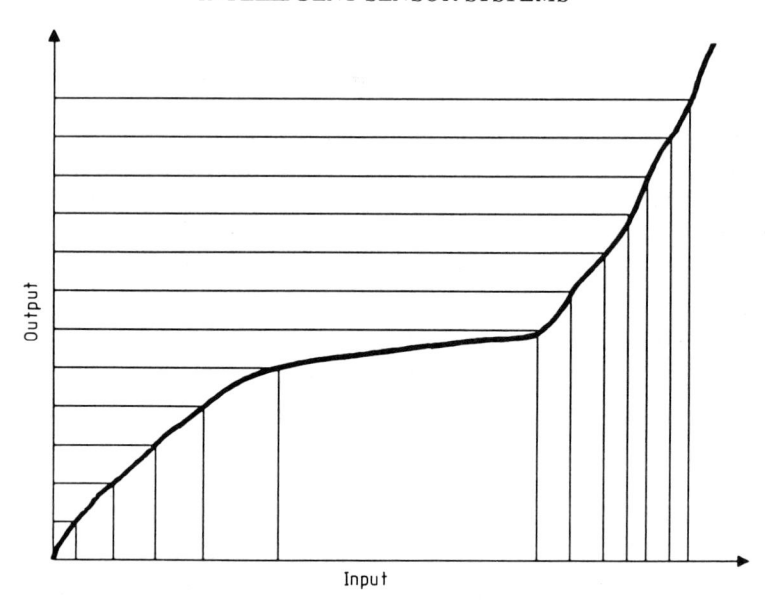

**Figure 3.3** Illustration of the effect of sensor characteristic on the system accuracy.

so the **minimum** slope of the characteristic is an important determinant of the **maximum** error in the target variable. Thus, while it is tempting to say that non-linearity is no longer a problem, gross non-linearities are still ruled out by virtue of the errors introduced.

**Noise** is a particularly difficult topic to encapsulate in a brief treatment such as this, and this difficulty arises from the fundamental definition of noise, which is **any unwanted signal**. Thus noise is defined by the nature of the **wanted** signal, and the methods available to combat it are determined by the nature of that signal. The term noise is also, however, used as shorthand for **random noise**, which is the difficult and all pervading form. Random noise is always present, if only because the universe is in a state of continuous agitation. The theory of random noise in physical devices is an important aspect of sensor technology, and is covered in more detail in § 4.4. Any designer of sensor systems should be familiar with its main results.

A well known example of a noise reduction method being determined by the nature of the wanted signal is the class of measurements in which it is known that that signal is repeated at known intervals, typically occurring when the test object can be subjected to a repeated stimulus. Here cumulative averaging (coherent averaging or box-car detection) can be used. Thus if $z_{rk}$ is the $r^{th}$ sample of the $k^{th}$ record then the improved averaged signal is given by the recursion

$$\bar{z}_{r,k} = \left(\frac{k-1}{k}\right)\bar{z}_{r,k-1} + \left(\frac{1}{k}\right)z_{r,k}; \qquad r = 1,2,3,\dots,m \qquad (3.2)$$

and the relative noise is reduced by a factor proportional to the square root of $k$. If the wanted signal is known to inhabit a particular restricted part of the spectrum then a filtering approach is indicated. A particular extension of this principle occurs in a class of measurements designed to detect whether a pre-defined signal is present in the noise. A classical example is the detection of the epileptic precursor in EEG (§ 6.4.1). Here a process known as matched filtering is used. A matched filter for a signal of shape $f(t)$ in the presence of white noise is one whose impulse response is the reversed form $f(T-t)$, where $T$ is the duration of the signal. It is easy to show that the output of the filter exhibits a positive peak whenever $f(t)$ occurs. For non-white noise the formula is more complicated (§ 6.3.4.6).

When the stimulus to the test object is disposable, more complex methods such as correlation may be exploited (§ 2.1.8). Since random noise is, by definition uncorrelated, correlation, or indeed any averaging process, will tend to reduce it. Time averaging of signals is a particular form of low-pass filter, which is often used, but it has to be used with care as its frequency domain implications are not entirely simple.

There is one particular type of noise which can cause great difficulties with primary sensors. This is $1/f$ noise, so-called because the noise amplitude per unit bandwidth is inversely proportional to frequency, and is largely believed to be associated with surface states in materials. $1/f$ noise can pose very substantial problems, especially when it is desired to measure down to zero frequency. By its very nature it is immune to time averaging techniques, and it is often difficult to distinguish from drift. The techniques of dealing with $1/f$ noise are somewhat specialized, but may be required in particular cases (Wilmshurst 1985).

**Parameter drift** is another defect which is invariably present to a greater or lesser extent. It may be confused with an unidentified cross-sensitivity, but strictly the term refers to slow changes in the nature of the lump of material that forms our primary sensor. The possible reasons for such changes may be manifold. The crystalline structure of materials may change owing to such processes as fatigue. Atoms of one material forming the lump may stray across the boundary with a different material by a process of diffusion. Surfaces may adsorb and absorb gases and vapours, oxidation being a particularly prevalent example. Other slow acting chemical changes may also take place. Thus our primary sensor is never quite the time-invariant system we think it should be.

The problem of dealing with parameter drift is similar to the noise problem in that the potential solution is conditioned by the nature of the wanted signal. It is a great help if we are not required to measure down to zero frequency, as slow acting offsets can be filtered out. This process appears in various forms (e.g. baseline correction).

Unfortunately parameter drift does not only affect the offset of the primary sensor: it may also affect the gain (figure 3.4). Where this form of the phenomenon is significant the only effective solution is periodic calibration, or preferably auto-calibration.

No primary sensor is sensitive to one physical variable only, and this fact gives rise to the important defect known as **cross-sensitivity**. By far the most important

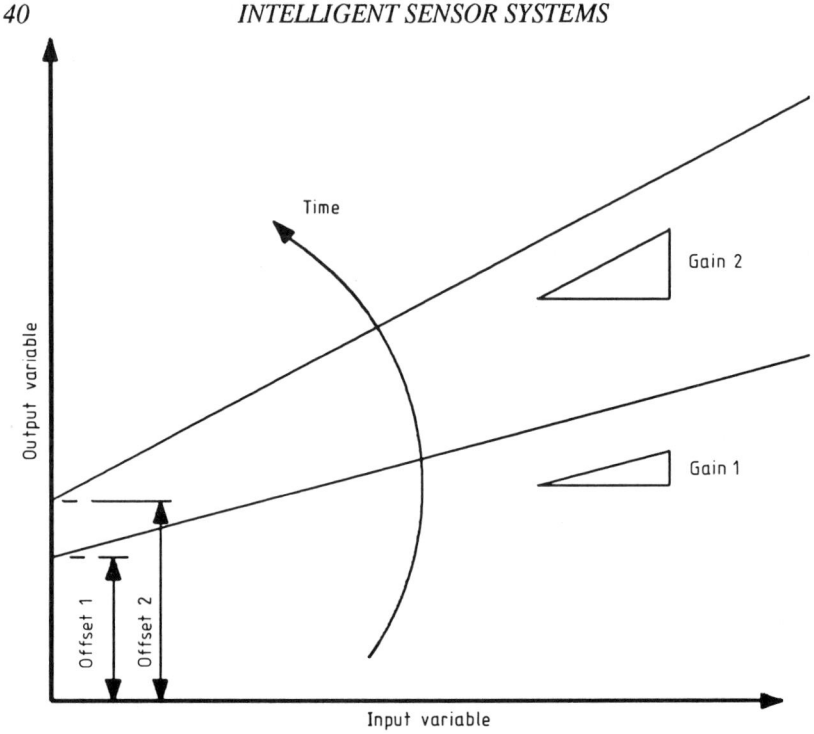

**Figure 3.4** Parameter drift can affect both offset and gain.

form of cross-sensitivity is with temperature. Virtually all physical processes are temperature dependent, and so are all our uncompensated primary sensors. In many ways temperature is a particularly complicated form of cross-sensitivity as it produces both variable and parametric changes. Consider a strain gauge (§ 3.3.1.2); thermoelectric effects may produce a change in the output variable (voltage), fortunately normally negligibly, but a change of temperature produces a change in the parameters. Furthermore this occurs not only as an offset but also as a gain change. The latter is generally much more difficult to compensate. In order to discuss methods of compensation for cross-sensitivity we need to survey the forms of compensation that are available.

Like any other set of ideas techniques of compensation can be classified in various ways. For the present purpose the authors have found it useful to divide these techniques into four classes of compensation (Brignell 1987b).

      1) Structural compensation
      2) Tailored compensation
      3) Monitored compensation
      4) Deductive compensation.

**Structural compensation** refers to the most traditional form of compensation found in sensors and it concerns the way the material forming the sensor is physically organized to maximize the sensitivity of the device to the target variable and to minimize response to all other physical variables. A good example is the precision load cell (§ 8.2). In this case not only the mechanical structure of the device is symmetrical, but so is the electrical structure (in the form of a Wheatstone bridge), and this illustrates the most fundamental manifestation of structural compensation, which is *design symmetry*. The idea behind design symmetry is that the target variable is arranged to produce a difference signal, while all other physical variables produce a common mode signal. Of course, besides cross-sensitivity, structural compensation is applied to the other defects enumerated above, but its application to the cross-sensitivity problem is particularly important.

Once structural compensation is applied at the design stage there is inevitably a residual defect for which it cannot cater, and this residue will vary between nominally identical sensors as they come off the production line. In the case of the precision load cell the strain gauges forming the bridge, their bonding to the metal structure and that structure itself will always reveal asymmetries which contribute to the residual error. The basic load cell as it comes off the production line will show a variation of output with temperature plus a sensitivity to eccentricity of the applied load. The former is traditionally dealt with by including an appropriate length of resistance wire of a given temperature coefficient in one arm of the bridge; the latter by filing pieces of metal off appropriate parts of the structure (formerly by hand and latterly by robot). As we shall see later the smart sensor approach is more likely to comprise the down loading of a distinctive set of coefficients into a ROM within the electronic sub-system (§ 6.4.3). All of these techniques which require action determined by the individual sensor and not the overall design come within the classification of **tailored compensation.**

Our third class of compensation techniques, **monitored compensation** is peculiar to the intelligent sensor approach. Thus if, as is generally the case, temperature sensitivity is a problem, the intelligent approach is to measure the temperature and use it to provide compensation computationally, either by reference to a model of the sensor or, preferably, by making use of data obtained from a calibration cycle. This principle can be applied to any other interfering physical variable. The tool for monitored compensation is the *sensor-within-a-sensor,* but when the cross-sensitivity problem is so severe that it becomes one of lack of *specificity* we have to appeal to the *sensor array* (§ 6.4.2)

**Deductive compensation** is a final class of compensation which is resorted to in special circumstances where for one reason or another the test object is not physically accessible. Examples of such objects would be the human brain, a nuclear reactor or the cylinder chamber of an internal combustion engine. Deductive compensation requires reference to a model, and because all models are imperfect it is only used as a last resort. In this form of compensation models of various degrees of sophistication may be employed.

As a very simple example of this technique consider a power transistor circuit. The electronic engineer designing such a circuit knows that a limiting factor is the junction temperature, but he has no means of measuring it. He knows that he can make an estimate of it from a knowledge of the external temperature and the power input, and would normally refer to the constraint of a hyperbola of constant power on the output $v/i$ characteristic. He also knows that under fast transient conditions he can transgress this barrier, thereby sub-consciously using a model of the device which is of the form of a low-pass filter. If he wanted to be really elaborate he could use a finite element model of the electrical and thermal fields in the material of the device, but the foundations of measurement data would probably be too flimsy to support such an imposing structure. This simple illustration epitomizes the problematic nature of such methods, but when nothing else is available they have to be employed.

## 3.3 SURVEY OF PRIMARY SENSING MECHANISMS

### 3.3.1 Mechanical

A full discussion of the numerous categories of mechanical sensors is beyond the scope of this book. The interested reader is advised to consult the existing texts which cover the complete field (Neubert 1963, Norton 1982, Doebelin 1990). However, there are a number of areas which need to be addressed; namely pressure, force, acceleration, displacement and flow.

#### 3.3.1.1 *Pressure*
Pressure is one of the most common physical parameters requiring measurement. For example meteorologists need to monitor atmospheric pressure, the automotive engineer may require a knowledge of inlet manifold pressure and the process control engineer may possibly wish to measure the pressure of compressed gases in pipelines. It is evident then that the range of pressure measurement likely to be encountered may cover low pressures (less than one millibar) up to extremely high pressures (thousands of bars). Different sensing mechanisms need to be employed for each category of low, mid and high range pressure measurement so we will limit our discussion to devices which perform in the range 1 to 5000 bar. Sensors for measurement of pressure outside this range are covered elsewhere (Norton 1982).

It is important at this stage to appreciate some of the terminology used in pressure measurement and to distinguish between absolute, gauge and differential pressures. Absolute pressure sensors measure pressure with reference to a zero pressure (i.e. a vacuum), whereas gauge pressure is defined as measurement with reference to atmospheric pressure. Differential pressure devices measure the difference between two pressures $P_1$ and $P_2$ (for example, across an orifice as described in § 3.3.1.5). The relationship between gauge and absolute pressure is:

$$\text{Absolute pressure} = \text{Gauge pressure} + \text{Atmospheric pressure} \quad (3.3)$$

***U-tube manometer***. The U-tube manometer has been one of the most common forms of pressure measurement device used in industry and provides a visual measurement of pressure to an operator. The various forms of the device can be

# PHYSICAL PRINCIPLES OF SENSING

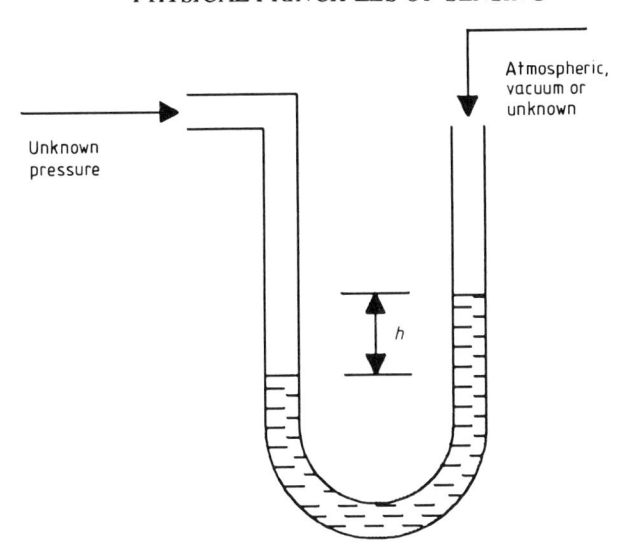

**Figure 3.5** U-tube manometer.

made to measure pressures in the range 1 to 7000 bar. The output is not usually in electrical form and thus would not be expected to be found in a control system or an intelligent instrument. The reason for its inclusion in this text is to illustrate the underlying principles of gauge, absolute and differential pressure measurement.

This device comprises a glass vessel in the shape of the letter U which contains a quantity of liquid and is shown in figure 3.5. The unknown pressure is applied at one end and the other may be exposed to a vacuum, ambient or another unknown pressure so that the device can be used to measure absolute, gauge or differential pressures respectively. A general expression for the pressure difference $(P_1 - P_2)$ is given by:

$$(P_1 - P_2) = h\rho g \tag{3.4}$$

where $\rho$ is the density of the liquid and $g$ is the local gravitational acceleration. The difference between the pressure heads $h$ can be read from a calibrated scale. The exact nature of the liquid depends upon the mode of operation of the manometer and the required pressure range. Water (actually coloured water for obvious reasons) is a convenient liquid to use in terms of cost but it is prone to evaporation and cannot be used to measure pressure of certain fluids which react or dissolve in water. Mercury is often used in these devices, particularly where high pressures are encountered.

***Bourdon tube.*** Bourdon tubes exist in a variety of forms and two are depicted in figure 3.6. In most cases the device consists of an oval cross-sectioned tube which is fixed at one end and free to move at the other. When a pressure is applied at the fixed end the oval cross-section becomes circular in nature and this results in a displacement at the free end.

The maximum displacement is proportional to the angle of arc through which the tube is bent, and with the C-type device, shown in figure 3.6 (a), this is usually less than 360°. However, C-type tubes are often used in the measurement of pressures up to 5000 bar and a tube radius of 25 mm and maximum displacement of 4 mm is typical. If greater resolution and sensitivity are required the spiral type device as shown in figure 3.6 (b) would be used, where the size of arc subtended is limited by the how many turns it is possible to implement easily and inexpensively. The displacement can be amplified mechanically via a gear quadrant/pinion system or by an electro-mechanical displacement sensor such as the LVDT discussed later.

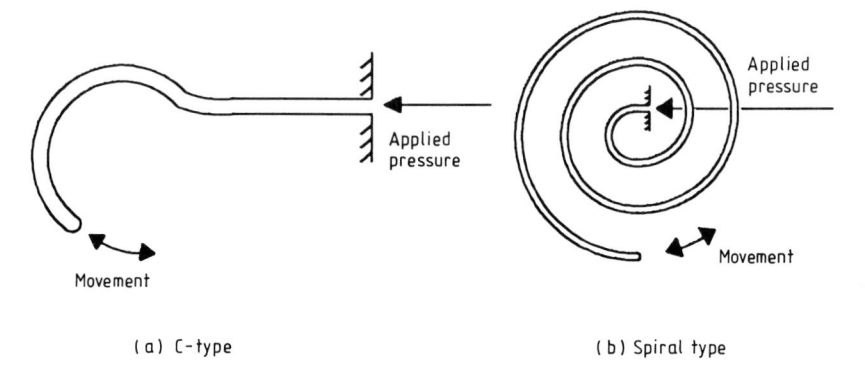

**Figure 3.6** Examples of Bourdon tube constructions.

*Pressure diaphragms.* An elastic circular diaphragm provides the basis for many modern pressure sensors. If a circular plate is clamped around its circumference and a pressure $P$ is applied to one side of the diaphragm then the plate will deflect in the middle by an amount $x$. This displacement can be measured with an appropriate sensor such as the LVDT (see § 3.3.1.2). However, in order to achieve a linear relationship between the applied pressure $P$ and the deflection $x$ it is essential to keep $x$ as small as possible. As a rule of thumb the deflection $x$ is usually designed to be less than half the diaphragm thickness $t$. The most common way to obtain an electrical output from the clamped diaphragm is by using resistance strain gauges.

A typical example of a pressure diaphragm is depicted in figure 3.7. The four strain gauges are mounted on one side of the diaphragm while the other side is exposed to the applied pressure. This basic arrangement can be found in many forms of diaphragm pressure sensor, comprising different diaphragm materials and utilizing a variety of enabling technologies for the gauges. For example, micromachined silicon devices have a physical size of only a few hundred μm (Blasquez *et al* 1989, Mallon *et al* 1990) and use doped silicon resistors as the strain gauges,

## PHYSICAL PRINCIPLES OF SENSING

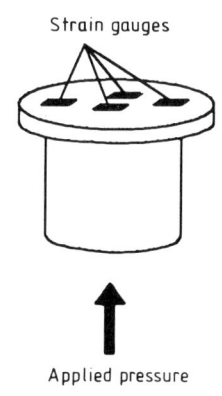

**Figure 3.7** An example of a diaphragm pressure sensor.

while a low cost device is based on the piezoresistive properties of thick-film resistors fabricated onto a ceramic, or steel, diaphragm (Catteneo *et al* 1980, Holford *et al* 1990). The most common form, however, consists of a stainless steel diaphragm with special-purpose metal foil strain gauges.

The strain distribution across the diaphragm is shown in figure 3.8. The radial and tangential (or hoop) components of strain, $\varepsilon_r$ and $\varepsilon_t$ respectively, are given by (Dalley and Riley 1978)

$$\varepsilon_t = \frac{3P(1-v^2)}{8Et^2}(R_o^2 - r^2) \qquad \varepsilon_r = \frac{3P(1-v^2)}{8Et^2}(R_o^2 - 3r^2) \qquad (3.5)$$

$$\varepsilon_t = \varepsilon_r = \frac{3PR_o^2(1-v^2)}{8t^2E}$$

$$\varepsilon_r = \frac{-3PR_o^2(1-v^2)}{4t^2E}$$

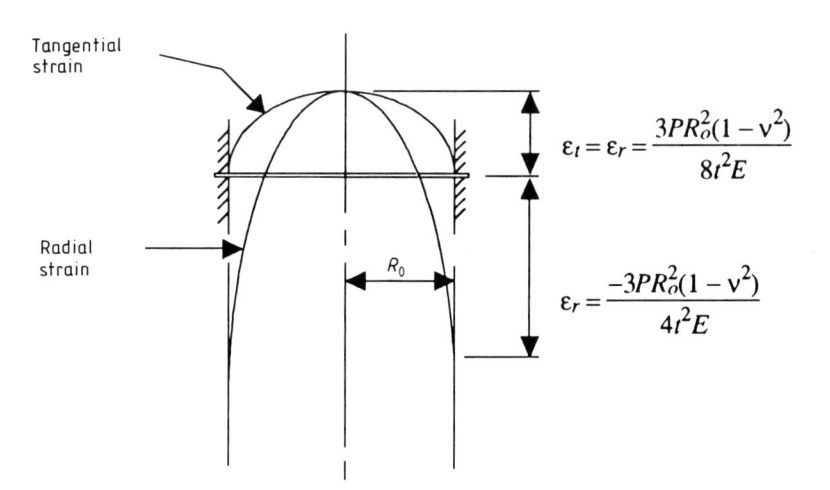

**Figure 3.8** The strain profile across the surface of a pressurized diaphragm.

where $P$ is the applied pressure, $t$ is the thickness of the diaphragm, $R_o$ is the radius of diaphragm, $r$ is the position parameter, $v$ is Poisson's ratio and $E$ is the modulus of elasticity of the diaphragm.

It is clear from the diagram that the tangential component is always positive and attains its maximum value at $r = 0$. The radial component is positive in a region close to the centre of the diaphragm but is negative at the periphery. By suitable positioning of the gauges a four-arm active bridge can be realized, allowing a high sensitivity and first-order temperature compensation.

Another form of pressure sensor diaphragm incorporates a number of concentric corrugations which increase the total effective area of the diaphragm. This arrangement is known as a corrugated diaphragm and provides a larger deflection, for a given pressure, than a flat diaphragm. It is usual to use this device in association with a displacement sensor to provide a compound pressure sensor.

### 3.3.1.2 *Displacement*

Displacement sensors provide the basis for many mechanical transducers for the measurement of force, pressure, velocity and acceleration. The displacement may be a translational motion, a rotational motion or a combination of both. The underlying principles for the measurement of these components are common, in combination with other primary sensing mechanisms, and in the following text we will restrict our discussion to linear displacement sensors.

**Resistive techniques.** A resistive potential divider (potentiometer) provides the basis for a simple displacement sensor. The device consists of a resistance element and a movable contact as shown in figure 3.9; the electrical circuit analogy is also presented. $R_t$ and $R_l$ represent the total resistance of the element and the load resistance respectively. If a voltage $V_s$ is applied across the element (length $x_t$) and the contact is attached to a body whose motion is being measured, then under no-load conditions a linear relationship will exist between $x_m$ and $V_o'$ (no-load output voltage); this can be expressed by

$$V_o' = \frac{x_m}{x_t} V_s .  \qquad (3.6)$$

In practice it is impossible to measure the voltage $V_o'$ without drawing extra current through the resistive element, and hence this loading effect must be appreciated by the instrument designer in order to minimize it by using a suitable buffer circuit.

Analysis of the circuit gives the following expression for the output voltage $V_o$

$$\frac{V_o}{V_s} = \left( \frac{x_t}{x_m} + \frac{R_t}{R_l}(1 - \frac{x_m}{x_t}) \right)^{-1} . \qquad (3.7)$$

## PHYSICAL PRINCIPLES OF SENSING

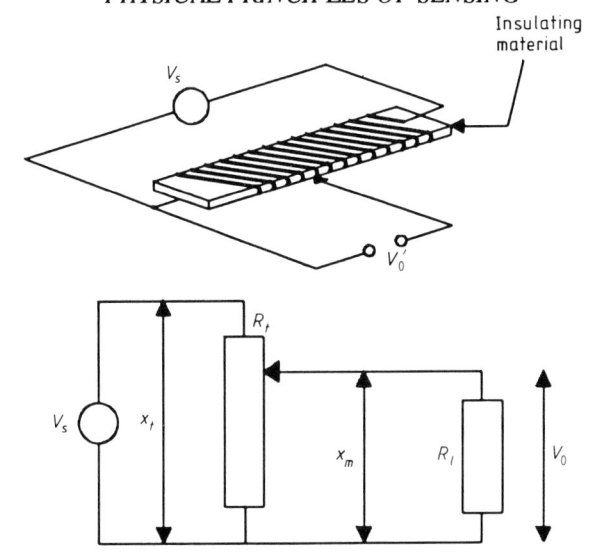

**Figure 3.9** A potentiometer displacement sensor and equivalent electrical circuit with load applied.

It is clear from the above equation that for finite values of $R_l$ the relationship between $V_o$ and $x_m$ is non-linear. For example, if $R_t/R_l = 1$ the maximum error could be as high as 12 % of full scale output (FSO). For a given value of $R_l$, $R_t$ is required to be sufficiently low to give a linear output. Increasing the supply voltage $V_s$ is another available option for achieving a high sensitivity with a low value of $R_t$ but care must be taken not to exceed the heat-dissipating capacity of the resistive element. It is clear then that there will always be a trade-off between linearity and sensitivity.

The resolution of the wire-wound potentiometer is limited by the number of turns per unit distance that is practically achievable. Such a device will also possess significant inductance, which will need to be accounted for if an AC excitation is to be used. A higher resolution can be achieved by using a carbon film, a cermet (ceramic-metallic) film, a metal film or a conductive plastic as the resistive element.

The main problem with potentiometers occurs at the point of contact between the wiper and the resistive track, which is particularly vulnerable to dirt and wear due to friction. High-speed motion of the slider can result in contact bounce thereby producing an intermittent output. The typical life expectancy of a potentiometer is of the order of 20 million full strokes, but this is dependent upon the nature of the material used for the resistive element.

***Strain gauges.*** Strain is a dimensionless geometric quantity which is related to the deformation of a body and is a consequence of the forces applied to it. For the purpose of this text we may loosely think of strain as being a relative change in the shape of a body. On this basis the fundamental principles of electrical resistance strain gauges are now introduced. The surface stress $\sigma$ is related to the surface strain $\varepsilon$ by the well-known expression $\sigma = \varepsilon E$, where $E$ is the modulus of elasticity of the material. The relationship between stress and strain was first observed by Robert Hooke in the 17th century. About one hundred years after the initial discovery Robert Young defined the constant of proportionality mathematically and it became known as Young's modulus. During the early 19th century Poisson found that strain applied in one direction produced a strain normal to the direction, proportional to it, and of opposite polarity, the constant of proportionality being Poisson's ratio, $\nu$.

In 1856 Lord Kelvin discovered the principle on which the electrical resistance strain gauge is based. He noticed that the resistance of various metal wires changed when strained. Furthermore, he observed that different wires changed by differing amounts when subjected to the same applied strain. The question which arises from this is whether the change in resistance is due to a dimensional effect or a change in resistivity with applied strain.

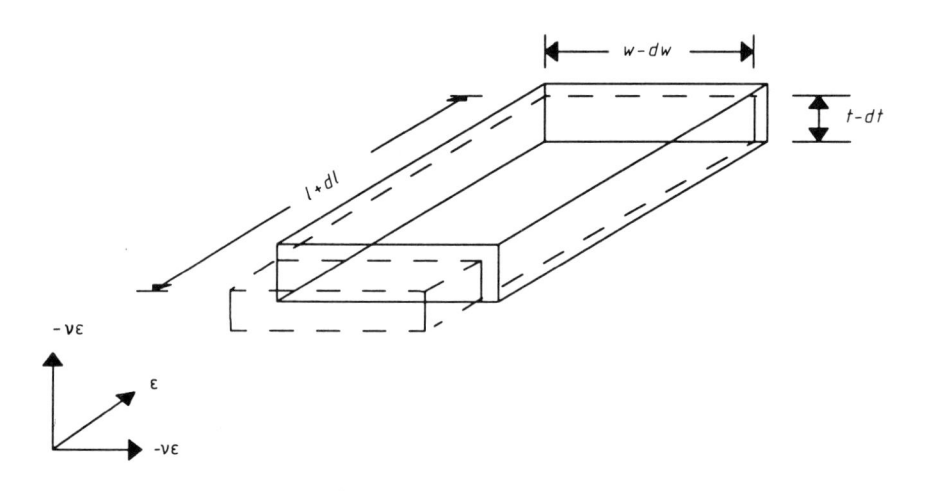

**Figure 3.10** Uniform model of a conductor subjected to an applied strain.

The resistance of a metallic alloy of length $l$, cross-sectional area $A$, and bulk resistivity $\rho$, is given by

$$R = \frac{\rho l}{A} \qquad (3.8)$$

logarithmic differentiation leads to

$$\frac{dR}{R} = \frac{d\rho}{\rho} + \frac{dl}{l} - \frac{dA}{A} \qquad (3.9)$$

## PHYSICAL PRINCIPLES OF SENSING

The term in $dA/A$ represents the fractional change in area due to a strain $\varepsilon$, along the axis of the body (i.e. $dl/l = \varepsilon$). The cross-sectional area is the product of the width $w$, and the thickness $t$, assuming uniform dimensions. Figure 3.10 shows how the dimensions change as a result of strain applied along the length $l$. As a consequence of Poisson's ratio $\nu$, the width and thickness will decrease by $dw$ and $dt$ respectively. Hence,

$$dw = -\nu w \varepsilon \text{ and } dt = -\nu t \varepsilon. \tag{3.10}$$

The cross-sectional area changes to $A'$ where

$$A' = (w-dw)(t-dt) = wt - 2\nu wt\varepsilon + \nu^2 wt\varepsilon^2 \tag{3.11}$$

Typically $\nu$ will lie between 0.2 and 0.3 and $\varepsilon$ will be of the order of a few micro-strain ($10^{-6}$), so the term in $\nu^2 wt\varepsilon^2$ is extremely small compared with the other terms in the equation and the change in area $dA$ is approximately,

$$dA = A' - A = -2\nu\varepsilon A \tag{3.12}$$

giving,

$$\frac{dR}{R} = \frac{d\rho}{\rho} + (1+2\nu)\varepsilon. \tag{3.13}$$

The sensitivity $S$ of the material to an applied strain is commonly called the **gauge factor** and is defined as

$$S = \frac{dR/R}{\varepsilon} \tag{3.14}$$

hence,

$$S = \frac{d\rho/\rho}{\varepsilon} + (1+2\nu). \tag{3.15}$$

Clearly the gauge factor comprises two distinct effects; the *piezoresistive* effect $(d\rho/\rho)/\varepsilon$ and the geometric effect $(1+2\nu)$. The common metal foil strain gauge has a gauge factor of around 2.1 typically, the geometric effect is dominant and contributes around 75 % of the overall sensitivity. Semiconductor strain gauges may have a gauge factor in the range $\pm$ 100 depending on the nature of the impurity dopant and its concentration. The piezoresistive effect is greater in semiconductive materials and the geometric effect contributes around 1% to the overall sensitivity. A relatively new class of strain gauge is based on the piezoresistive properties of

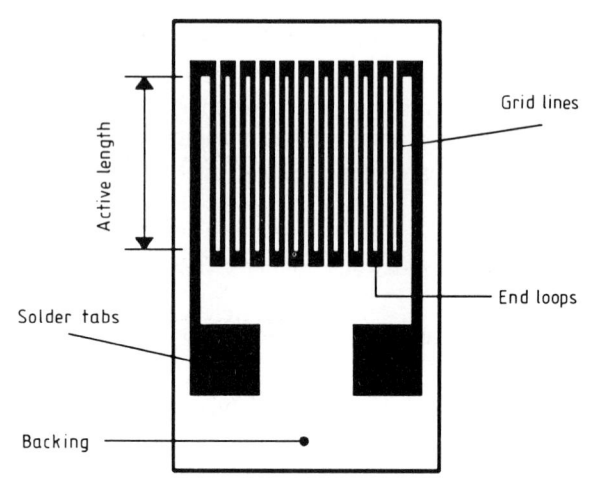

**Figure 3.11** Metal foil strain gauge.

thick-film resistors (Prudenziati *et al* 1981, White 1988). These devices have a gauge factor of around 10 and the piezoresistive effect dominates the geometric contribution.

An example of a metal foil gauge is shown in figure 3.11. The grid comprises a number of parallel lines which give a total gauge length exceeding the active length thereby providing a greater resistance. Most foil gauges have a resistance of 120 or 350 ohms. The backing material is a polyimide film: being both tough and flexible this provides suitable insulation to isolate the gauge electrically from the specimen to which it is attached. The end loops are elongated in order to minimize the effect of transverse sensitivity.

*Capacitive techniques.* The capacitance, in farads, of a parallel plate capacitor of plate area $A$, separation $d$, is given by

$$C = \frac{\varepsilon_o \varepsilon_r A}{d} \quad (3.16)$$

assuming there are no fringing effects. $\varepsilon_o$ is the permittivity of free-space and $\varepsilon_r$ is the relative permittivity of the material between the plates. It is clear then that the capacitance varies in proportion to either the change in area $A$ or $\varepsilon_r$, and is inversely proportional to the distance $d$. These simple relationships form the basis for many capacitive sensors. A moving dielectric device is depicted in figure 3.12 (a). The two electrodes are separated by a fixed distance and have a constant area of overlap. Any motion of the movable dielectric, (permittivity $\varepsilon_r$) will cause a change in capacitance between the plates. Figures 3.12 (b) and (c) show variable area and variable separation capacitive displacement sensors respectively. In general there are limitations on the separation $d$ with these simple parallel plate

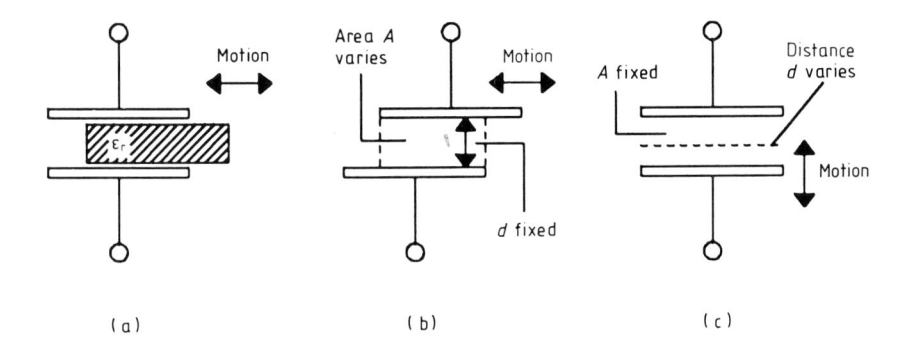

**Figure 3.12** Examples of capacitive displacement sensors. (a) moving dielectric, (b) variable area, (c) variable separation.

capacitor arrangements. The maximum separation is restricted by the fringing effects which cause the capacitance to vary in a non-linear manner. Most practical variable separation capacitance sensors operate over a range of a few millimetres. Electric breakdown is not a constraint at normal electronic voltages owing to Paschen's law. The arrangement shown in figure 3.13 provides a linear characteristic for larger displacements. This device is a differential capacitance sensor and consists of three electrodes $X, Y$ and $Z$ where $X$ and $Y$ are fixed and $Z$ is movable. The capacitance between $X$ and $Z$ is denoted by $C_2$ and that between $Y$ and $Z$ by $C_1$. When $Z$ is midway between $X$ and $Y$ $C_1$ is equal to $C_2$. If $Z$ is displaced by an amount $x$ towards plate $X$ then,

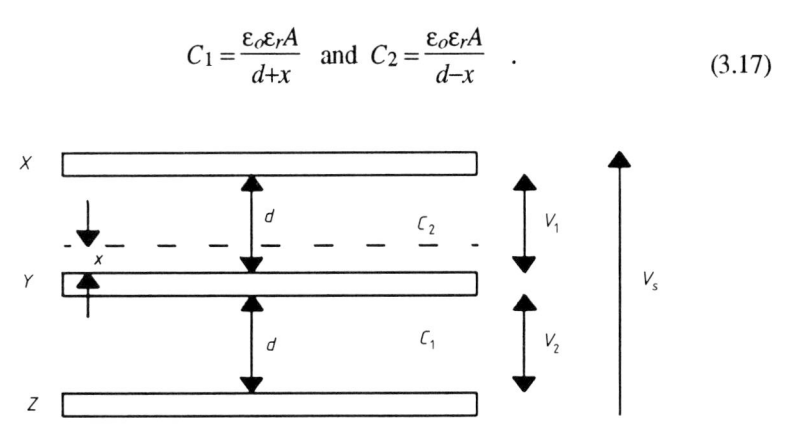

$$C_1 = \frac{\varepsilon_0 \varepsilon_r A}{d+x} \quad \text{and} \quad C_2 = \frac{\varepsilon_0 \varepsilon_r A}{d-x} \quad . \tag{3.17}$$

**Figure 3.13** Principle of a differential capacitance sensor.

The differential output voltage ($V_1$–$V_2$) for a supply voltage $V_s$ across X and Y is given by,

$$(V_1 - V_2) = V_s \left( \frac{C_2}{C_1 + C_2} - \frac{C_1}{C_1 + C_2} \right) = \frac{V_s x}{d} . \qquad (3.18)$$

Hence there is a linear relationship between output voltage and the displacement $x$, and this remains linear over a range of about $|x| < d$. This technique has been successfully used for the measurement of displacements in the range $10^{-11}$ m up to $10^{-2}$ m.

Stray capacitance from cables in parallel with the sensor will affect the linearity and sensitivity of the device. This is particularly noticeable if very small changes in capacitance are being sensed (< 1 fF). Cross sensitivities to temperature and humidity are prevalent with most types of capacitive sensor, especially if an air gap is used as the dielectric medium.

*Inductive techniques.* A commonly used inductive displacement sensor is the linear variable differential transformer (LVDT). This sensor comprises a single primary coil and two secondary coils connected in series opposition as shown in figure 3.14. The object whose motion is to be measured is attached to the ferromagnetic core. As the core is displaced, the mutual inductance between the primary and the two secondaries changes giving rise to the following induced emf's in the secondary coils,

$$V_1 = k_1 \sin(\omega t - \varphi) \ ; \ V_2 = k_2 \sin(\omega t - \varphi) \qquad (3.19)$$

where $k_1$ and $k_2$ depend on the amount of coupling between the respective secondary and primary coils, and therefore also on the position of the ferromagnetic core. If we consider the case where the core is in a central position the coupling between the respective secondaries will be equal (ideally). Owing to the series opposition arrangement of the secondaries the output voltage $V_o$ will be the difference between $V_1$ and $V_2$. Hence $k_1 = k_2 = k$, and $V_1 = V_2 = k \sin(\omega t - \varphi)$ giving $V_o = 0$. In practice, due to small variations in construction, a slight residual voltage remains when the core is in its central position. Suppose now that the core is moved upwards by a distance $x$. The coupling between respective secondaries and the primary is different, say $k_1 = k_a$ and $k_2 = k_b$, the output voltage is now

$$V_o = (k_a - k_b) \sin(\omega t - \varphi) . \qquad (3.20)$$

If the core is displaced by an equal amount $x$, in the opposite direction from the null position, the values of $k_1$ and $k_2$ will become $k_2 = k_a$, $k_1 = k_b$ and the output voltage is now $V_o = (k_b - k_a) \sin(\omega t - \varphi)$. Alternatively this can be rewritten as

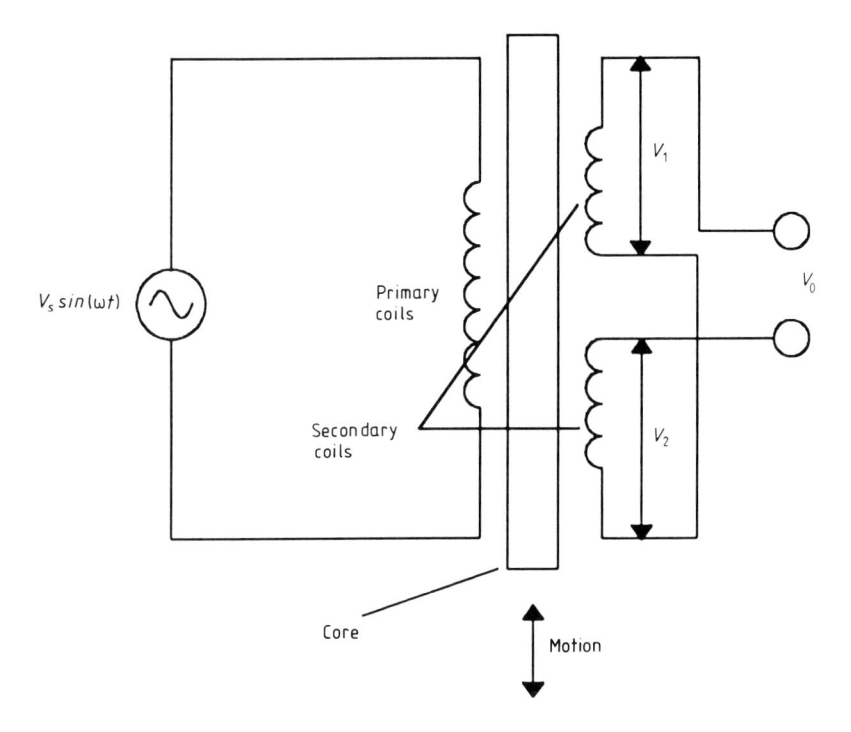

**Figure 3.14** Circuit diagram of a typical LVDT.

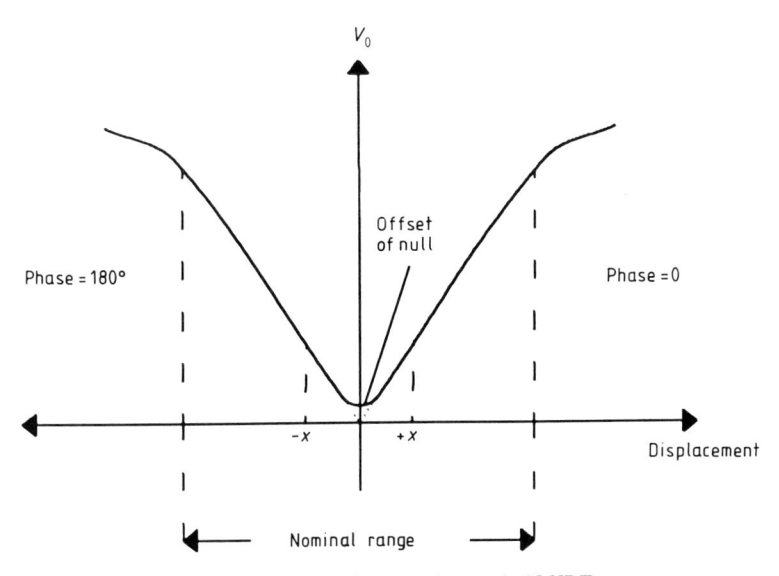

**Figure 3.15** Output voltage versus core displacement for a typical LVDT.

$$V_o = (k_a - k_b) \sin(\omega t + (\pi - \varphi)) \ . \tag{3.21}$$

The magnitude of the output voltage for equal displacements around the null point is identical. The only difference being a phase shift of 180° as shown in figure 3.15, the inherent offset at the null position is also indicated. It is possible to obtain LVDTs which contain a phase sensitive detector connected to the output which provides a positive or negative voltage relating to a displacement of $+x$ or $-x$ respectively. Furthermore, devices exist which allow a DC excitation and produce a DC output (typically ±15 V input, ±10 V output). LVDTs usually cover the displacement range from ±100 µm to ±100 mm. The only moving part is the central core which moves only in the air gap between the coils. There is virtually no friction or wear during operation and long life expectancies are quoted by the manufacturers, around 200 years typically.

***Optical techniques.*** Another way for measuring displacement is the so-called Moiré-fringe method. If two fine line gratings are arranged so that they lie at a slight angle to one another light and dark fringes will be formed when light is directed through the pair. When one plate is displaced in the transverse direction an increased movement in the fringes is seen in the perpendicular direction. A suitable radiant sensor like a photocell (see § 3.3.5) can be placed on opposite sides of the light source to sense the variations in light intensity. The number of peaks detected will indicate the magnitude (but not the direction) of the displacement. If another photocell is positioned such that there is a 90° phase shift in light intensity with respect to the other photodetector then the direction of motion can be sensed. With a grid density of around 100 lines/mm it is possible to measure displacements down to a few µm.

### 3.3.1.3 *Force*

Force on a mass arises from two properties of matter, inertia and gravitation. These yield two possible methods of measuring the force. The first technique is to measure the acceleration of a body of known mass and the second method involves a force-balance approach whereby the unknown force is balanced against the gravitational force acting on a known mass. The former method is of little practical value as in most cases the unknown force is part of a system which is not free to accelerate. Force-balance systems are encountered in a variety of applications including shop weighing scales and domestic weighing appliances. The design may be a simple beam balance or a complex integration of mechanical gearing and pendulums.

The above techniques describe the more traditional ways of force measurement and have now virtually become redundant as most modern-day force sensors are based on elastic sensing elements which essentially convert an input force into an output displacement according to another physical law which obeys the following equation

$$F = kx \tag{3.22}$$

## PHYSICAL PRINCIPLES OF SENSING

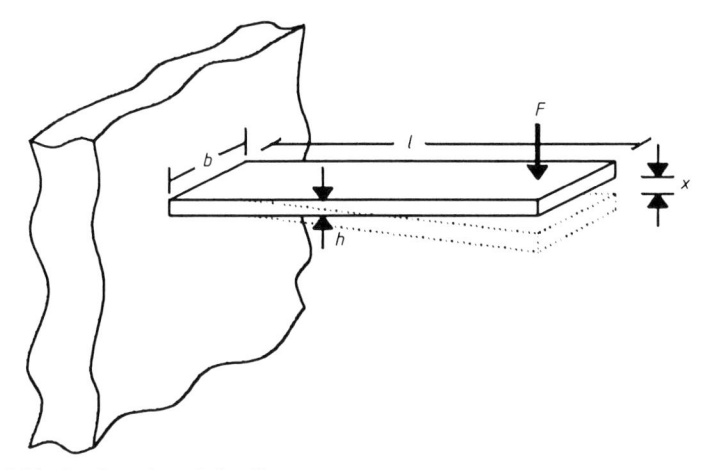

**Figure 3.16** Cantilever beam in bending.

where $k$ is a constant and $x$ is the displacement produced. A simple spring balance is an example of a device utilizing this basic principle.

Another simple example of an elastic sensing element is a bending beam of which there are three main types; (i) the encastré (built-in beam), (ii) the simply supported beam, (iii) the cantilever beam (shown in figure 3.16).

The displacement $x$ at the free end of the cantilever beam is related to the force $F$ by the following expression

$$x = \frac{4Fl^3}{Ebh^3} \tag{3.23}$$

where $l$ is the length of the beam, $b$ is the breadth, $h$ the thickness and $E$ is the modulus of elasticity. The above relationship is only valid for small deflections at the free end. The magnitude of the force $F$ is proportional to the displacement $x$ which can be measured using a suitable technique.

The surface strain $\varepsilon$ at a distance $d$ from the applied force is given by

$$\varepsilon = \frac{6Fd}{Ebh^2} \, . \tag{3.24}$$

Again, if the force $F$ causes a small deflection at the free end of the cantilever a linear relationship exists between the surface strain $\varepsilon$ and the force $F$. Many different techniques could be used to measure the strain but most practical force sensors use a form of resistive strain gauge element. Industrial force sensors utilizing the elastic element technique are sometimes referred to as load cells. Most commercial load cells are based on a bending beam principle. One major drawback of using a single beam in load cell design is that the surface strain is often a function of the location of the load positioning, so for a given force the output will vary with load eccentricity. A common way of minimizing this effect is to use an arrangement similar to that shown in figure 3.17. This structure is known as a coupled-double-beam and is essentially two beams coupled together at each end. The strain gauges are sited over the areas of minimum thickness. The dumb-bell cut-out arrangement provides areas of maximum surface strain whilst keeping the overall structure rigid. Unlike the single beam cantilever load cell the coupled-double-beam does not have a linear strain profile across its length, the structure bends in an S-shape thereby giving rise to areas of tensile and compressive strains which enable the strain gauge to be easily configured in a four-arm active bridge arrangement. The coupled-double-beam is less sensitive to eccentrically applied loads than the cantilever beam and is also more resistant to side loads.

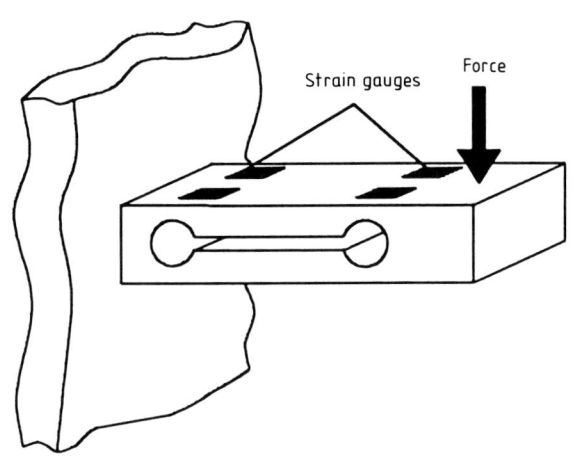

**Figure 3.17** A coupled-double-beam load cell.

### 3.3.1.4 *Acceleration*

The class of sensor used for the measurement of acceleration is termed an accelerometer. By definition acceleration is the second derivative of displacement so initially it may appear that the obvious way to obtain a reading of acceleration is to differentiate the output of a displacement sensor first to obtain velocity and again to obtain acceleration. However, this technique is seldom used in practical devices as the process of differentiation always amplifies the high frequency noise in the measurement system (§ 4.5.4). Nearly all accelerometers which are commercially available are based on a spring-mass-damped system as shown in figure 3.18. The spring and the damper are assumed to be rigidly fastened to the accelerating body. The spring has stiffness $k$ and the damper has a constant denoted by $\lambda$. When an acceleration $a$ is applied to the body at a time $t = 0$, the resultant force is given by Newton's second law

## PHYSICAL PRINCIPLES OF SENSING

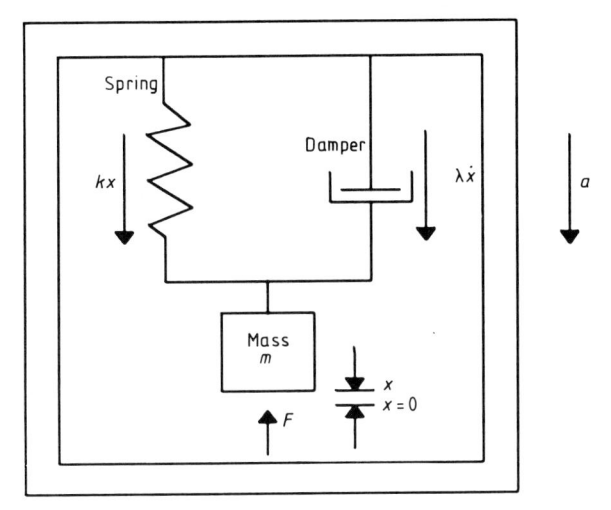

**Figure 3.18** A spring-mass-damped accelerometer.

$$F = m\ddot{x} + \lambda\dot{x} + kx = ma \qquad (3.25)$$

In the above equation $\ddot{x}$ represents the acceleration of the mass relative to the casing and can be different to the acceleration $a$ of the casing itself. It is clear that this is a typical second-order linear differential equation an can be rewritten as

$$\frac{m d^2 x}{k\, dt^2} + \frac{\lambda\, dx}{k\, dt} + x = \frac{ma}{k}. \qquad (3.26)$$

The undamped natural frequency is

$$\omega_n = \left(\frac{k}{m}\right)^{1/2} \qquad (3.27)$$

and the damping ratio is given by

$$\zeta = \frac{\lambda}{2\sqrt{(km)}}. \qquad (3.28)$$

A common form of accelerometer consists of a cantilever beam (spring) which has a seismic mass attached to its free end. Damping is usually air or a viscous liquid. Figure 3.19. shows such a system. The undamped natural frequency of the accelerometer is (Neubert 1963)

$$\omega_o = \frac{12.2\pi}{l}\left(\frac{nh}{\varepsilon}\right)^{1/2} \tag{3.29}$$

where $l$ is the length of the cantilever, $h$ is the thickness, $\varepsilon$ is the surface strain at the fixed end and $n$ is the range of the sensor in units of '$g$'. It is usual to position strain gauges on each side of the cantilever, in a full bridge configuration, to obtain an output voltage signal. The mass $m$, required to produce an acceleration value of $ng$ is approximately

$$m = \frac{\varepsilon E b h^2}{6nl} \tag{3.30}$$

where $E$ is the modulus of elasticity of the beam material and $b$ is the breadth. A high natural frequency and low mass are usual requirements for an accelerometer thus indicating that $\omega_o/m$ is large. We have

$$\frac{\omega_o}{m} \propto \frac{1}{bE}\left(\frac{n}{\varepsilon h}\right)^{3/2} \tag{3.31}$$

which suggests that the modulus, thickness and width of the beam should be kept low to achieve a high value of $\omega_o/m$.

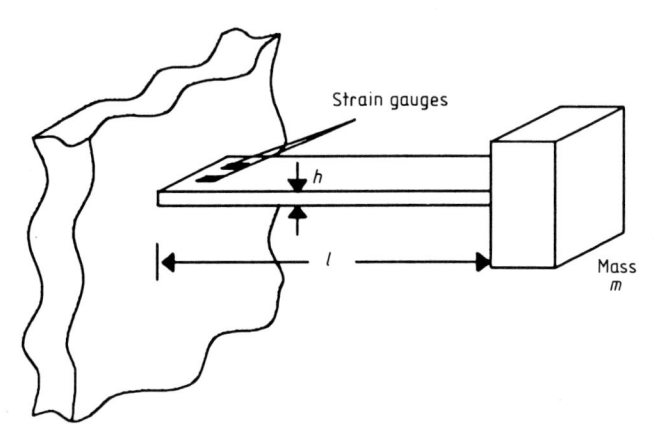

**Figure 3.19** A cantilever beam-based accelerometer.

Another type of accelerometer utilizes the piezoelectric effect exhibited in certain materials like quartz and various ceramics. When a force is applied to a piezoelectric material its dimensions change and a potential difference can be measured across opposite faces. The effect is reversible in that if a voltage is applied across opposite faces the physical dimensions change. Some accelerometer designs use the piezoelectric material both as the spring and the damper within the sensor thus enabling a small light-weight device to be made. One major drawback with piezoelectric accelerometers is that they are not suitable for measuring constant or slowly varying accelerations.

Reluctive accelerometers using a LVDT fixed between two parallel cantilever beams also exist. With these devices the ferromagnetic core acts as the seismic mass. All acceleration sensors exhibit a cross-sensitivity to accelerations at right angles to the sensing axis, this error is expressed as a percentage of full-scale output (FSO), a typical LVDT-based device having a cross-sensitivity of around 0.5 % FSO, whilst a cantilever beam device would posses an error of about 1 % to a transverse acceleration.

By integrating the output of an accelerometer velocity can be determined. The process of integration tends to reduce high frequency noise in the measurement system thereby providing a suitable technique to sense the velocity of a body.

### 3.3.1.5 *Flow*

Flow measurement is of particular importance to the process industries. A wide variety of sensors are available to measure the flow rate of solids, liquids, gases and slurries. Flow rate can be concerned with the velocity of flow (m/s), the volumetric flow rate ($m^3$/s) or the mass flow rate (kg/s). Mass flow rate is usually applicable to the measurement of the flow rate of solids and a common example is that found in the weighing industry utilizing a belt-weighing technique to measure the mass of a particulate composition of solid material (coal, shingle or the like) over a fixed length of the belt. The mass flow rate is obtained by the product of the velocity of the belt and the mass per unit length of material. The mass flow rate of gases, liquids and slurries can also be determined using instruments such as the Coriolis flowmeter and these are described elsewhere (Loxton and Pope 1985). A full discussion of the numerous methods for the measurement of the flow rate of solids, liquids gaseous and non-homogeneous material would occupy several volumes of text. It is the intention of this sub-section to describe some of the underlying principles and give examples of flow rate sensor which are applicable to intelligent sensor systems.

*Differential pressure devices.* When a constriction is present in a fluid-carrying pipe a pressure difference is created on either side of the obstruction. The square root of the differential pressure is proportional to the volumetric flow rate of the fluid. Three specific devices which utilize this technique are the Venturi tube, the orifice plate and the Pitot tube. These are illustrated in figures 3.20 and 3.21. Figure 3.20 (a) shows a general characteristic of the pressure distribution preceding and after the point of constriction.

In order to derive a mathematical relationship between the pressure drop and the volumetric flow rate Bernoulli's theorem can be used. In the following analysis it will be assumed that; (i) the fluid of interest is in steady state flow, (ii) there are no frictional forces between the fluid and the pipe walls or in the fluid itself, (iii) there is no heat losses or gains due to heat transfer between the fluid and its surroundings, (iv) the fluid is incompressible.

From Bernoulli's theorem we can say that at two positions in the fluid the following equation relating to the conservation of total energy is valid

$$\frac{P_1}{\rho_1} + \frac{v_1^2}{2} + gz_1 = \frac{P_2}{\rho_2} + \frac{v_2^2}{2} + gz_2 \qquad (3.32)$$

where $P_1$ and $P_2$ are the pressures, $v_1$ and $v_2$ are the velocities at the respective positions, $z_1$ and $z_2$ are the elevations above a datum and $g$ is the gravitational acceleration. From assumption (iv) above we can say that $\rho_1 = \rho_2 = \rho$, the fluid density. Furthermore if we assume that the pipe is horizontal $z_1 = z_2$ and the equation reduces to

$$\frac{(v_2^2 - v_1^2)}{2} = \frac{P_1 - P_2}{\rho}. \qquad (3.33)$$

Defining $Q$ as the volume flow rate and using assumption (i) we get

$$Q = Q_1 = Q_2 \text{ where } Q_1 = A_1 v_1 \text{ and } Q_2 = A_2 v_2 \qquad (3.34)$$

$A_1$ and $A_2$ are the cross-sectional areas of the fluid at the two respective positions. Equation (3.31) now becomes

$$Q = \frac{A_2}{\left[1 - (A_2/A_1)^2\right]^{1/2}} \left[\frac{2(P_1 - P_2)}{\rho}\right]^{1/2}. \qquad (3.35)$$

The above equation is the ideal theoretical equation for the volumetric flow rate. However, this equation does not hold true in practical situations. One reason for this is that the assumption of frictionless flow is not obeyed in practice. For smooth pipes with turbulent flow the friction loss is small and constant but non-zero. The Reynolds number, $N_R$ specifies the ratio of turbulence forces to viscous forces forces and is given by $N_R = vD\rho/\eta$, where $D$ is the internal diameter of the pipe and $\eta$ the fluid viscosity. Reynolds numbers less than 2000 indicate laminar flow, whereas turbulent flow becomes more dominant when the Reynolds number exceeds 2000. Complete turbulence exists for a Reynolds number greater than $10^4$. Another reason for the non-applicability of equation (3.33) is that $A_1$ and $A_2$, the cross-sectional areas of the fluid at the two stations are difficult to measure and may also change with the rate of flow. A modified equation allows the actual flow rate of an incompressible fluid to be determined by

$$Q' = \frac{C_D A_2'}{\left[1 - (A_2'/A_1')^2\right]^{1/2}} \left[\frac{2(P_1-P_2)}{\rho}\right]^{1/2} \tag{3.36}$$

where $A_1'$ and $A_2'$ are the cross-sectional areas of the pipe at the points of observation and $C_D$ is a constant known as the discharge coefficient which is a function of the Reynolds number $N_R$ and the difference between pipe and flow diameters. The value of $C_D$ can be determined experimentally but tables of values are available for given fluid and pipe diameters.

The orifice plate is an example of a common type of flowmeter found in industry and is essentially a metal disc with a concentric hole in it as depicted in figure 3.20 (b). It is easy to produce, low cost and available in a wide range of sizes. One drawback with this device is that over a period of time the edges of the hole tend to wear away and also small particles or even bubbles in the fluid can build-up behind the hole and reduce its diameter. The discharge coefficient may also vary with time.

The Venturi tube is a precision engineered device and is a very expensive instrument with a high accuracy. The device is shown in figure 3.20 (c). One potential drawback with all pressure difference flowmeters is that a permanent

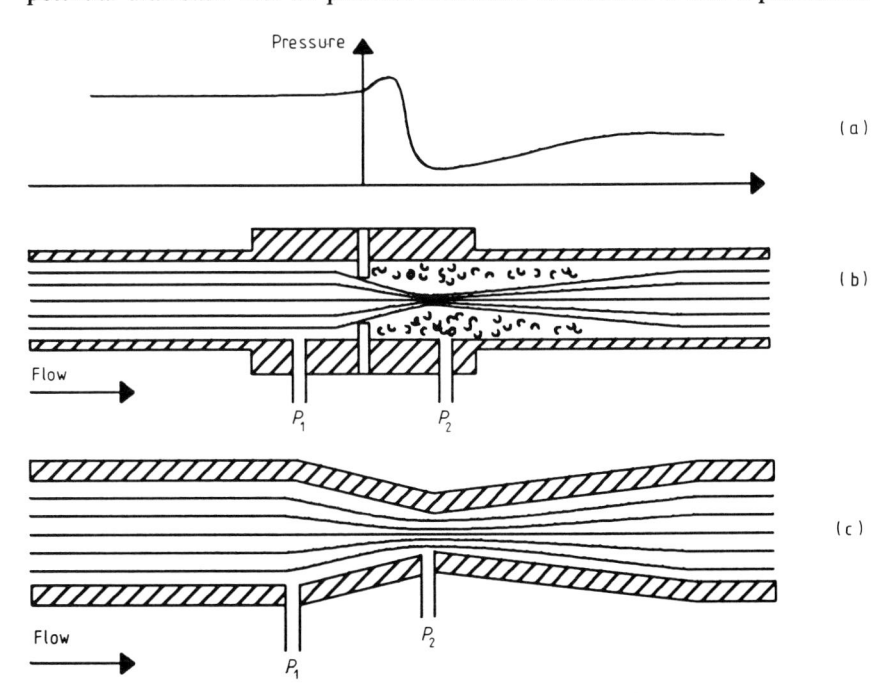

**Figure 3.20** Differential pressure-based flowmeters. (a) pressure profile along the length of pipe, (b) an orifice plate flowmeter, (c) a Venturi flowmeter.

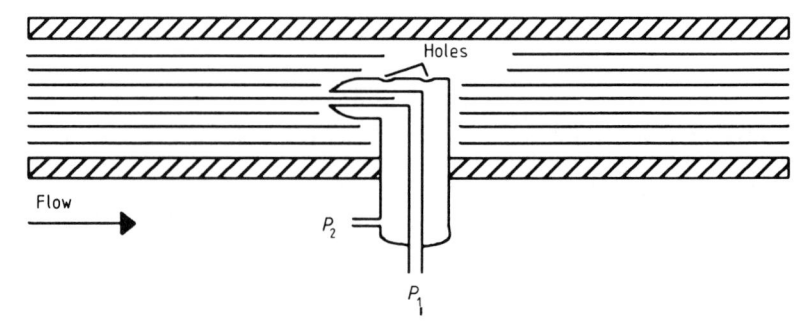

**Figure 3.21** The Pitot tube flowmeter.

pressure loss occurs downstream of the point of obstruction as illustrated in figure 3.20 (a). However, this effect is minimized with the Venturi tube due to its geometric design.

The Pitot tube ( figure 3.21 ) consists of two concentric tubes installed into the fluid-carrying pipeline. The inner tube has its open end facing the oncoming fluid and the outer tube has a closed end with a number of holes around its circumference. When the fluid impinges on the impact hole it is brought to rest and therefore has no kinetic energy only potential energy. If $v$ is the mean flow velocity, $P_1$ is the pressure at the entrance to the tube and $P_2$ is the 'static' pressure then from Bernoulli's equation we have

$$v = \left[\frac{2(P_1-P_2)}{\rho}\right]^{1/2}. \tag{3.37}$$

In practice not all the incident fluid on the inner tube will be brought to rest so a correction factor is sometimes needed to accommodate for this. It is usual to measure the pressure difference $(P_1-P_2)$ with an inclined tube manometer and the volume flow rate can be determined by multiplying $v$ by $A$, the cross-sectional area of the inner tube. If the flow velocity profile is not uniform across the diameter of the pipeline then a number of Pitot tubes can be used to infer the mean volumetric flow rate.

*Variable-area devices.* This class of flowmeter gives a visual indication of the volume flow rate of gases and liquids. The device comprises a vertical tube with a tapered bore in which a 'float' attains a different equilibrium position at different flow rates. The float itself usually resembles a truncated cone which presents an obstruction to the flow in the vertical direction. The word 'float' is perhaps an unsuitable choice, as the density of the so-called float must be greater than that of the fluid whose flow rate is being measured. As the flow rate is varied the vertical position of the float alters in order to present a different orifice area at each flow rate. Typically, the accuracy of a variable-area device would be expected to be around ± 2 % FSO but fully calibrated versions can be as high as 0.2 % FSO.

*Turbine flowmeters.* A turbine flowmeter consists of a multi-bladed rotor mounted in a pipe along a parallel axis to the fluid flow. When the fluid impinges on the blades they rotate at a rate proportional to the flow rate of the fluid. A ferromagnetic material is usually chosen for the blade construction thus providing the basis for a variable-reluctance tachogenerator by the addition of a permanent magnet and coil inside the housing. The voltage pulse induced in the coil as each blade passes can be fed into a suitable pulse counter and the pulse frequency may thus be determined. A high degree of accuracy is achievable with these devices ( ±0.1%) but they are prone to wear of the turbine wheel if there is any particulate matter in the fluid.

*Vortex-shedding flowmeters.* When a fluid flows past an unstreamlined obstacle (bluff body) the flow cannot follow the contours of the body downstream and a number of vortices are created in the low-pressure area behind the body. The frequency of vortex shedding is proportional to the flow rate of the fluid. The vortices can be detected by a number of differing techniques based on capacitive, thermal, magnetic and ultrasonic principles.

*Electromagnetic flowmeters.* If the fluid is electrically conductive and moving through a pipe of diameter $D$ with average velocity $\bar{v}$ then by Faraday's law of electromagnetic induction the induced voltage $E$ between two electrodes across the diameter of the pipe is

$$E = BD\bar{v} \qquad (3.38)$$

where $B$ is the magnetic flux density applied orthogonally to both the fluid flow and the electrodes. The volume flow rate can be easily determined by multiplying the velocity $\bar{v}$ by the cross-sectional area of the tube. The magnetic field coils are usually energized by an alternating current thus necessitating a suitable demodulation of the induced voltage. An AC magnetic field is preferred to DC in order to prevent polarization around the electrodes and minimize the effects of other types of electrical interference like stray DC voltages. The fluid-carrying pipe is fitted with an insulating liner material such as PTFE (polytetrafluoroethylene) or polyurethane. This type of device has the advantage of not imparting any restriction to the fluid flow and therefore causing no pressure loss associated with the measurement.

*Hot-wire anemometer.* The hot-wire anemometer is a popular device for the measurement of the velocity of air and other gaseous fluids, some designs can also be used with liquids. The 'wire' can be either a metallic wire (sometimes a film) or another appropriate resistive sensing element like a thermistor. The element is incorporated into a Wheatstone bridge arrangement. Two possible circuit configu-

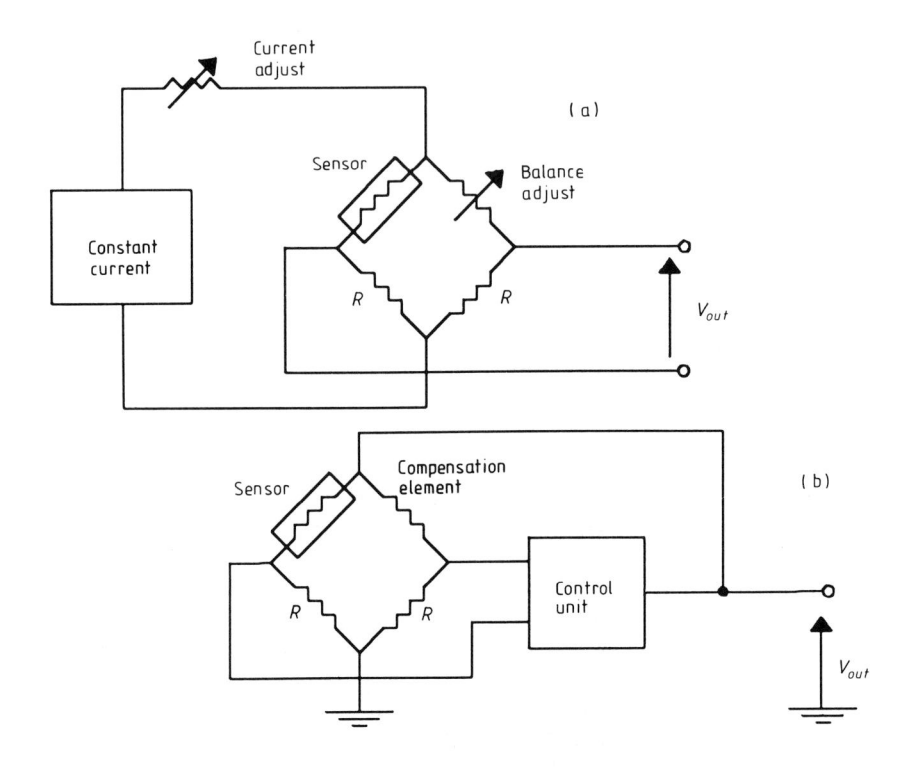

**Figure 3.22** Circuit configurations for a hot-wire anemometer: (a) constant current, (b) constant temperature.

rations are illustrated in figure 3.22. The circuit for a constant current anemometer shows that the sensor is heated from a constant current supply. The current can be adjusted to heat the sensing element to the required temperature for a given application and the bridge is balanced by the balance adjust resistor. As the fluid velocity changes the bridge becomes unbalanced thereby leading to a change in output voltage, the magnitude of which is proportional to the square root of the fluid velocity ( Bentley 1988).

The constant temperature anemometer has a compensation resistor in the adjacent arm of the bridge to the sensing element. The two components have similar temperature coefficients of resistance thus allowing the effects of variation in ambient operating temperature to be minimized as the two are sited in close proximity. When the fluid velocity changes the resistance of the sensing element changes causing the bridge to become unbalanced. The output from the bridge is fed into a control unit which amplifies the error signal and adjusts the bridge supply voltage until the bridge is balanced. The excitation voltage is thus a function of the fluid velocity.

## PHYSICAL PRINCIPLES OF SENSING

***Laser Doppler flowmeter.*** The laser Doppler flowmeter provides a direct reading of flow velocity of a liquid in a transparent tube. Light from the laser is focused by an optical lens onto a point in the flow. Any particles or bubbles present in the fluid will cause a Doppler shift of the scattered light which is received at a photodetector mounted near the laser. The frequency shift is proportional to the fluid velocity.

Ultrasonic flowmeters based on the same principle are also available and can be used on opaque or transparent pipes as ultrasound propagates well through most solid materials. Doppler flowmeters (§ 6.3.4.8) are mounted externally to the fluid-carrying pipe and being a non-contact method of flow measurement they do not impose any permanent pressure loss in the fluid. These devices find greatest use for measuring the flow rate of difficult fluids like corrosive liquids and slurries.

***Cross-correlation flowmeter.*** Cross-correlation techniques were discussed earlier in § 2.1.8 and these can be applied to flow measurement. If the fluid in question has a detectable random variable (e.g. density, temperature, conductivity) then by placing two appropriate sensors at a fixed distance along the fluid-bearing pipe it is possible to determine the volume flow rate from a knowledge of the mean flow velocity. Suppose the two random signals are $x(t)$ and $y(t)$ and these represent the signals from the upstream and downstream sensors respectively. If the $x(t)$ signal is delayed by a time $\tau$ then the cross-correlation function $R_{xy}$ is given by

$$R_{xy}(\tau) = \lim_{T \to \infty} \frac{1}{2T} \int_{-T}^{+T} x(t-\tau)y(t) \, dt = \overline{x(t-\tau)y(t)} \qquad (3.39)$$

where $T$ is the observation time. $\tau$ can be varied until $R_{xy}(\tau)$ is a maximum. This optimum value $\tau_{opt}$ can be used to calculate the mean flow velocity $\overline{v}$ by knowing $\tau_{opt} = L/\overline{v}$, where $L$ is the distance between the sensors. Some form of digital signal processing (computer) is an essential requirement of this instrument and therefore it is a good example of an intelligent sensor.

### 3.3.2 Thermal

In any discussion relating to thermal quantities it is important to emphasis the difference between heat and temperature. The temperature of a body is a measure of the average kinetic energy of the molecules and is the potential for heat flow. Heat is energy and its flow is due to temperature differences between a system and its surroundings or between two systems (see table 3.1). Heat transfer can occur by one or more of the following:

Conduction i.e., diffusion through solids or stagnant fluids.
Convection i.e., the motion of a fluid.
Radiation i.e., by electromagnetic waves (requires no medium).

For the purpose of this text we will discuss thermal sensors based on thermoelectric effects, thermoresistive effects, and properties of *p-n* junctions. Other methods such as thermal expansion (clinical thermometer, bimetallic strip) or resonant frequency change (quartz thermometer) are covered elsewhere (Bentley 1985, Morris 1988, Doebelin 1990).

### 3.3.2.1 *Thermoelectric effects*
The three main thermoelectric effects being exploited in the measurement of thermal quantities are:

i) **The Seebeck effect**: This is observed when two dissimilar conductors (or semiconductors) are joined together at one point and a temperature difference exists between the joined and open ends. An open circuit voltage can be measured between the open ends (see figure 3.23)

ii) **The Peltier effect**: When a current passes through the junction of two different conductors, heat can be either absorbed or emitted at the junction depending on the direction of the current flow.

iii) **The Thomson effect**: This occurs when the temperature gradient exists along a single conductor and a current is passed through the conductor. Heat is absorbed or produced in addition to the release of Joule heat.

The Thomson effect is small in comparison to the other two and has found little use in practical applications.

The Peltier effect, unlike Joule heating, is a reversible effect. If heat is generated when current flows in one direction, the same quantity of heat is absorbed when the direction of the current is reversed. When suitable materials are chosen for example, it is possible to have a drop of water boil or freeze at the junction simply by reversing the current. Several years ago this effect attracted great interest as it was believed that domestic appliances could be manufactured which could be used as heaters in winter and coolers in summer, simply by reversing the switch. However, it was soon realized that the efficiency of Peltier devices is much less than conventional products. Peltier elements are not yet widely used except in special measurement instruments like the dew-point humidity sensor.

One of the most widely used temperature sensors today is based on the Seebeck effect and is referred to as a **thermocouple** and is shown in figure 3.23. The magnitude of the thermoelectric potential produced by a thermocouple is dependent on the properties of the two conductors and the temperature difference between the joined and unjoined ends. A general expression for the thermal potential is

$$\Delta V = A_1(T_1-T_2) + A_2(T_1^2-T_2^2) + A_3(T_1^3-T_2^3) + \ldots + A_n(T_1^n-T_2^n) \qquad (3.40)$$

where $A_1, \ldots, A_n$ are constants and $T_1, T_2$ are the temperatures at the joined and non-joined ends respectively. In practice certain combinations of conductor materials are chosen so that the higher order terms in $(A_2T_1^2, A_2T_2^2)$, $(A_3T_1^3, A_3T_2^3)$,

$(A_nT_1^n, A_nT_2^n)$, are negligible and the thermal potential to temperature relationship is approximately

$$\Delta V = A_1(T_1 - T_2) . \qquad (3.41)$$

The value of $A_1$ is dependent on the difference between the Seebeck coefficients of the materials used. Standard tables of values are available which list the thermal emf and hot junction temperature for various combinations of conductor materials. Most tables assuming a reference temperature of 0 °C at the open ends of the thermocouple. Metal alloys like alumel (Ni/Mn/Al/Si), constantin (Cu/Ni) and platinum/rhodium are often used in thermocouples and specific combinations are known by an internationally recognized type letter, for example type $K$ is chromel-alumel, type $E$ is chromel-constantin.

In many applications it is often not possible to maintain the reference temperature at 0 °C. Provision is made to locate the reference junction within a controlled environment at a temperature greater than zero. Correction of this can be made by using the law of intermediate temperatures which states that

$$V_{(T_1,T_o)} = V_{(T_1,T_r)} + V_{(T_r,T_o)} \qquad (3.42)$$

where $V_{(T_1,T_o)}$ is the thermal emf with the temperature difference $(T_1-T_o)$, $V_{(T_1,T_r)}$ is the emf produced with the junction at temperatures $T_1$ and $T_r$, and $V_{(T_r,T_o)}$ is the thermal emf observed when the temperature difference is $(T_r - T_o)$. $T_o$ is 0 °C, $T_r$ is the non-zero reference temperature and $T_1$ is the temperature being measured.

It is usual to have the measuring instrument positioned some distance away from the actual thermocouple, hence two more junctions are created at the thermocouple/connecting lead interface. If these compensation leads are manufactured

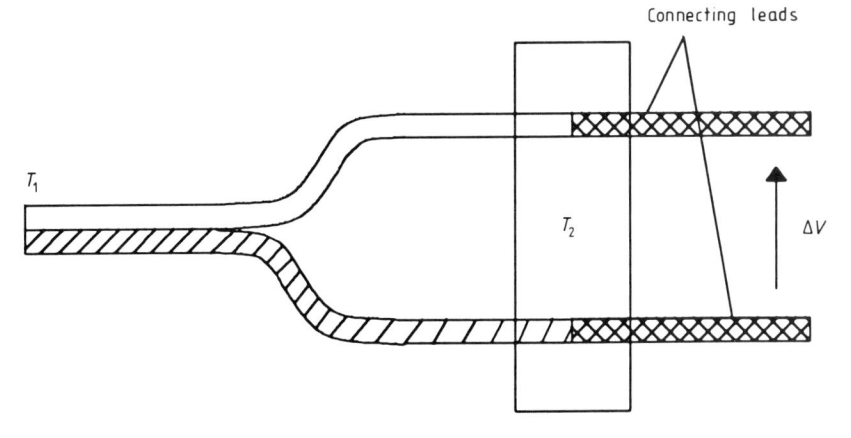

Figure 3.23 Joining two dissimilar metals together to make a thermocouple.

from similar conductor material to the respective thermocouple materials then the thermal emf's generated at the junctions will be negligible. However, the effect of the actual connection to the instrument still needs to be considered, as two more junctions are thus created because the metals are usually dissimilar. The law of intermediate metals indicates that the resultant thermal emf will not be affected by the inclusion of another junction provided that it is at the same temperature as the original junction between which the new metal was inserted.

Several thermocouples can be placed in series so that the reference junctions are at the same temperature and the coupled junctions are all exposed to the temperature being measured. This device is then referred to as a **thermopile**. The effect of having $n$ thermocouples increases the measurement sensitivity by a factor of $n$ and a typical device comprising 25 thermocouples has a resolution of around 0.001 °C.

Thermocouples are delicate sensors which are prone to errors arising from contamination and induced strain around the sensing junction. For these reasons it is usual to protect the thermocouple in a sheath. However, the addition of such protection may increase the time response of the system.

### 3.3.2.2 *Thermoresistive sensors*

Materials which exhibit a change in resistance in accordance with temperature are sometimes referred to as **thermoresistive**. If the material is a metal then the device is a resistance thermometer whereas a semiconductor material is often known as a thermistor. Both types of device are passive in nature and thus require an external supply voltage. A DC bridge is normally used to measure the resistance change. The bridge excitation voltage should be chosen carefully in order to achieve a high measurement sensibility but impose no self-heating effects in the resistive elements.

***Resistance thermometers:*** The variation in resistance with temperature for a metal is conventionally represented by a power series

$$R = R_o (1 + \alpha_1 T + \alpha_2 T^2 + \alpha_3 T^3 + ..... + \alpha_n T^n) \ . \tag{3.43}$$

This is clearly a nonlinear relationship but in practice metals like platinum, nickel and copper have a linear response over a limited temperature range (typically between -200 °C to +500 °C). The higher order terms ($\geq T^2$) in the above equation are therefore negligible and it reduces to

$$R = R_o (1 + \alpha T) \tag{3.44}$$

where $\alpha$ (ppm/°C) is the temperature coefficient of resistance (TCR) and $R_o$ is the resistance at a reference temperature (usually 0 °C). Platinum resistance thermometers (PRTs) would typically possess a TCR of around 4000 ppm/°C. Such sensors are very expensive but have a good linearity and chemical inertness. It is common to find PRTs comprising a platinum film rather than the bulk metal and these have a similar linearity and sensitivity to the bulk but are less expensive to fabricate (Reynolds and Norton 1985).

**Thermistors:** These are thermally sensitive resistors comprising a semiconductor material which is usually a sintered composite of a ceramic and a metallic oxide such as manganese, cobalt, copper or iron. They are purchased in the form of discs rods or beads. The resistance versus temperature characteristic is non-linear and is of the form

$$R = R_o \, exp\{\beta(1/T - 1/T_o)\} \quad (3.45)$$

where $R$ and $R_o$ are the resistances at temperatures $T$ and $T_o$ respectively and $\beta$ is a characteristic constant of the semiconductor material. By differentiating the above equation it is clear that

$$\frac{dR/R}{dT} = \alpha = -\frac{\beta}{T^2} \quad (3.46)$$

showing that the TCR ($\alpha$) is a function of temperature. The TCR has negative polarity and here the device is referred to as a NTC (negative temperature coefficient) thermistor. Some sintered oxides like BaTiO exhibit a positive temperature coefficient and are known as PTC thermistors. The manufacturing tolerance of absolute value is much wider for thermistors ($\pm 10 \%$) than metal resistance elements ($\pm 1 \%$).

### 3.3.2.3 *p-n junction temperature sensors*

**Diodes:** If a small, constant current flows through a silicon diode in the forward direction the forward voltage varies as a function of temperature. This relationship remains linear between 20 K and 300 K. A typical temperature coefficient would be 22 mV/ K, but this can vary between different devices due to the fabrication process (Middlehoek and Audet 1989). Germanium diodes also exhibit a similar temperature coefficient and both provide a low-cost, simple method of temperature measurement. However, owing to their poor interchangeability their use in broad industrial applications is limited and transistors are more commonly employed, as demonstrated in the following section.

**Transistors:** Bipolar transistors can be used as temperature sensors by exploitation of the $I_c$ - $V_{be}$ characteristics. The following approximate equation demonstrates the relationship between the forward bias collector current and temperature

$$I_c = I_{cs} \exp\left(\frac{qV_{be}}{kT}\right) \tag{3.47}$$

where $I_{cs}$ is the reverse saturation current, $k$ is Boltzmann's constant, $T$ is the absolute temperature, $V_{be}$ is the base-emitter voltage and $q$ is the electron charge. One possible circuit configuration for four matched transistors is shown in figure 3.24. Here the pair of matched transistors at the top act as a current mirror for the pair below. $T_3$ and $T_4$ have equal currents but different emitter areas. An expression for the output current $I_o$, is given by

$$I_o = \frac{2kT}{qR_o} \log_e \left(\frac{A_3}{A_4}\right) \tag{3.48}$$

where $A_3$ and $A_4$ are the emitter areas of $T_3$ and $T_4$ respectively, $A_3/A_4 \neq 1$. Hence the output current varies in proportion to the absolute temperature. The R5590 KH is an example of such an integrated circuit temperature sensor operating on a range 220 K to 420 K. By trimming the value of $R_o$ a nominal temperature coefficient of 1 µA/ K is achievable.

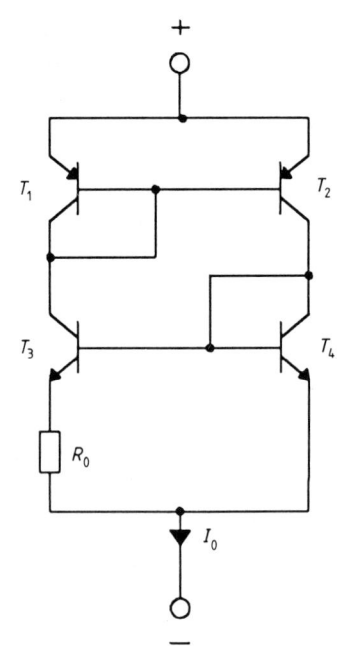

**Figure 3.24** A PTAT (Proportional to Absolute Temperature) circuit provides one way of using transistors to measure temperature.

### 3.3.3 Chemical

Modern day requirements for environmental monitoring have led to a rapid increase in research activity in the area of chemical sensors. Indeed, another book in this series is dedicated to current development in solid state chemical sensors (Mosley and Tofield 1987). Essentially there are two fundamental types of analysis for which chemical sensors are required; **qualitative** analysis is used to determine which elements are present in a sample and **quantitative** analysis reveals the amount of one or more given elements contained in the sample. The recent interest in gas analysis for example, in areas such as monitoring stack emissions in power stations requires quantitative analysis and measurements on exhausts gases in motor vehicles may also need quantitative analysis. However, in order to obtain the maximum amount of information both techniques are used simultaneously and this will usually require use of a computer.

This section will illustrate some of the more commonly used chemical sensors and in view of the recent demands for low-cost, miniature, disposable devices and because of limited space in this text, a coverage of solid state sensors is given in preference to the more traditional approaches.

#### 3.3.3.1 *Chemiresistors and chemicapacitors*

A popular form of chemical sensing element comprises a planar electrode pattern onto which an organic material is deposited. This device is sometimes called a chemiresistor. An example is in figure 3.25, here the sensor is fabricated using thick-film technology but similar devices can be made using thin-film or silicon technology. An organic material (a metal substituted phthalocyanine in the example) is deposited onto the interdigitated electrode pattern and this exhibits a change in conductance in the presence of a gas. The organic semiconductive material has a very low conductance and hence changes in resistivity are difficult

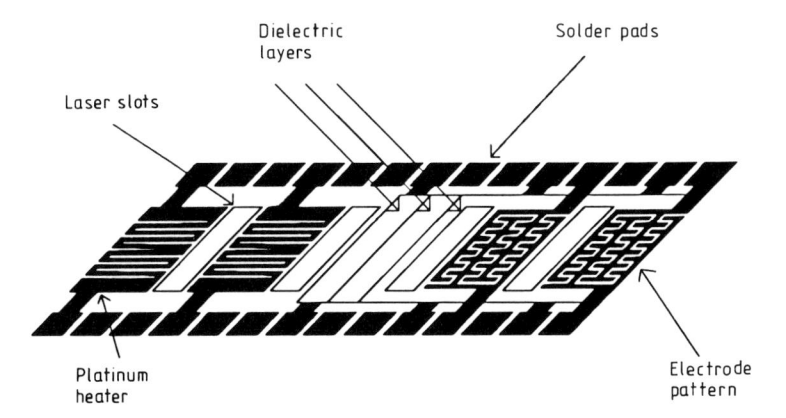

**Figure 3.25** An example of an array of chemiresistors fabricated using thick-film techniques. The slots are cut by a laser and help to isolate thermally each sensor site.

to measure, but by increasing the number of electrodes in the interdigitated electrode arrangement the changes become more measurable. The chemiresistor depicted in the diagram also has a platinum heating element which is separated from the electrode pattern by an insulating layer (low permittivity dielectric material). By supplying current to the heating element the localized sensor site can be heated to promote the chemical reaction between the organic layer and the sample gas. If the resistance of the heating element is monitored the temperature at the sensor site can be inferred (Brignell *et al* 1988). This type of sensor has been shown to be extremely responsive to low concentrations of various gases down to parts per billion (ppb) levels. The sensitivity to other gases is significant. If multicomponent analysis is required an array of chemiresistors can be constructed in which each element has a different reactive organic layer. Elaborate pattern recognition techniques are thus required to establish quantitative analysis of the mixture, and this is usually performed in software. A multi-element chemiresistor array provides a good example of an intelligent sensor system (§ 6.4.2, § 8.4.1).

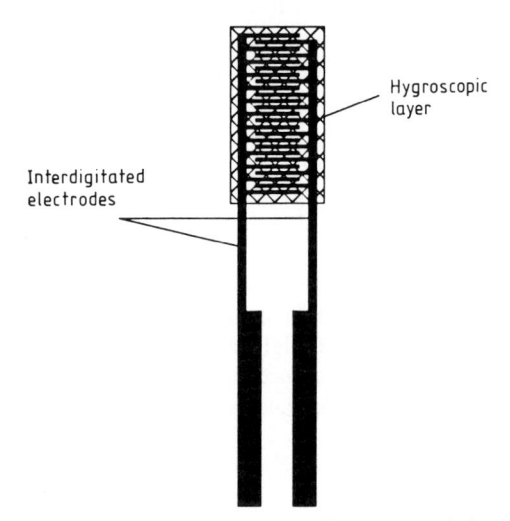

**Figure 3.26** A chemicapacitor used as a humidity sensor. The hygroscopic layer changes its dielectric constant in the presence of moisture.

Figure 3.26 shows an example of a chemicapacitor. It is similar in nature to a chemiresistor the main difference being that the change in capacitance between the electrodes is measured as opposed to the resistance change. The gas sensitive film is again deposited onto the electrodes. A simple humidity sensor can be easily made by depositing a hygroscopic dielectric material. The presence of moisture can be detected by monitoring the DC capacitance change between the electrodes.

## PHYSICAL PRINCIPLES OF SENSING

### 3.3.3.2 Ion - Sensitive Field Effect Transistors (ISFETs)

Probably the most extensively investigated solid state chemical sensor is the ISFET. This device is essentially a MOSFET without the metal gate electrode; a schematic cross-section is depicted in figure 3.27.

The gate insulator oxide is exposed to the sample solution, (an electrolyte). Silicon dioxide is commonly used as the gate insulator material. A more sophisticated version of an ISFET is a device known as a CHEMFET which has an ion-permeable layer on the gate oxide thus allowing a broader application field and increasing the specificity. The purpose of the reference electrode is to establish an influential electrochemical potential, and a silver chloride electrode is often used. The electrode is positioned at a distance from the silicon substrate and can be thought of as representing the gate of the device.

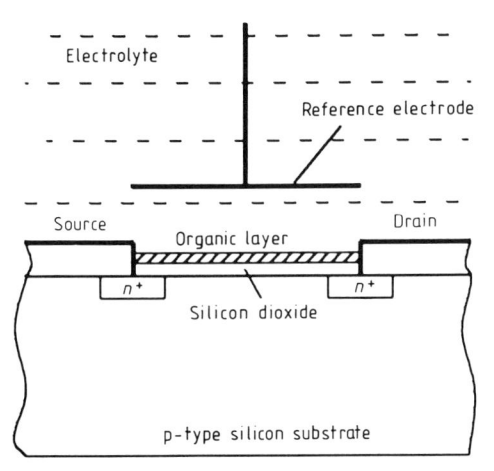

**Figure 3.27** Cross-section of a CHEMFET.

If a suitable gate to source bias voltage ($V_{GS}$) and drain to source voltage ($V_{DS}$) is applied to the ISFET and the gate insulator exposed to an electrolyte, then an electrochemical potential is developed at the solution-gate insulator interface and this affects the drain to source conductance. The drain current can be taken as a measure of this conductance. The electrochemical potential is a function of the concentration or, more correctly, the activities of the ionic entities in solution and the output drain current is thus associated with the electrolyte concentration. ISFETs are often used for measurement of the pH value of solutions (hydrogen ion activity) by using silicon nitride as the gate insulator. pH ISFETs typically operate over a pH range from 2-10 and have a fast response time of around 100 ms.

The addition of an ion-selective membrane to form a CHEMFET was originally believed to provide a good basis for the generation of a large family of chemical sensors exhibiting a wide selectivity and much research has been conducted on the use of CHEMFETs for biomedical applications. However, the number of commercially available sensors does not reflect the efforts of current research activity.

Some reasons why ISFETs have only found restricted usage are due to reliability problems associated with the ion-membrane deposition, unpredictable drift arising from fundamental characteristics such as $1/f$ noise, poor long term stability and the need for the addition of the reference electrode. The area of ISFET research is still popular, as indicated by the number of scientific papers and conferences on this subject (Bergveld 1970, 1985) and many researchers still believe that ISFETs will be developed to reach, and maybe surpass, the initial expectations conceived in the 1970s.

### 3.3.3.3 *Chromatography*

Chromatographs are used for both qualitative and quantitative analysis of gaseous or liquid mixtures. A schematic diagram of a gas chromatograph is show in figure 3.28. The sample is injected into a stream of inert carrier gas like helium or nitrogen which carries it through the separation column; a long, thin tube containing an adsorption material (packing). Substances with a high affinity to the packing tend to travel through the separation column at a slower rate than substances with a low affinity. The compounds exit from the separator in an identifiable sequence and pass through a detector (katharometer) which gives an output voltage signal proportional to the abundance of each compound. The elution products and carrier gas are then either vented or collected.

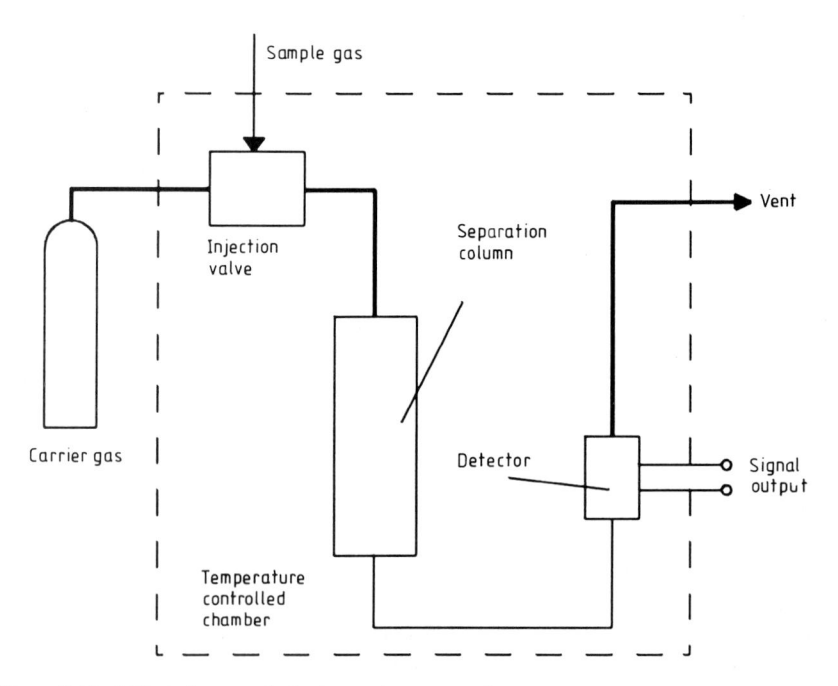

**Figure 3.28** A block diagram of a basic gas chromatograph.

Conventional gas chromatographs are not portable devices but research by Terry *et al* (1979) has attempted to fabricate a complete device onto a 5 cm silicon layer in order to achieve portability. The reported results appear to be promising, exhibiting high resolution and fast time response.

Chromatography techniques may also be used for analysing liquids. The primary difference between liquid and gas chromatography is the absence of a carrier gas. A dilute solution of the sample is passed through a column packed with solid particles during which time the components in the solution separate and thus give rise to variations in the travel time of different components. The effluent is passed through a detector such as a refractive index detector which produces an output signal indicating the difference between the refractive index of the sample and that of a reference liquid. Hence the constituent components can be determined by observation of the associated chromatograph (essentially a plot of detection output against time).

### 3.3.3.4 *Polarography*

Polarography is an electroanalytical technique for the determination of chemical species in gaseous or liquid surroundings. A classical approach uses an electrochemical cell with a 'dropping' mercury microelectrode as the cathode and a mercury pool electrode as the anode. Some electrochemical cells use precious metals (Pt, Au) or metal films as the electrodes and the more general term *voltammetry* is given to the analytical process. Microelectrodes are used in order to minimize the surface area and hence achieve the maximum diffusion current from the solution into the so-called *double-layer* at the electrode surface. This is a region where the chemical reaction predominately takes place. A typical electrochemical cell might comprise the electrolyte together with three electrodes the working electrode $W$, reference electrode $R$ and the counter electrode $C$. The reference electrode is used as a probe to monitor the potential generated at the working electrode relative to its own potential. Ideally the reference electrode draws no current.

A voltage is applied between the counter and working electrode and the resulting current is measured in the working electrical lead. The chemical reactions between the solution and electrodes is a complicated process, as is the exact nature of the double-layer. A more comprehensive approach to electrochemical cells is presented elsewhere (Kissinger and Heinman 1984) and the interested reader is advised to consult the many existing texts on electroanalytical chemistry for a more detailed discussion of the physio-chemical behaviour of this type of sensor.

Figure 3.29 shows a schematic diagram of a three-electrode electrochemical cell incorporated into a circuit known as a potentiostat. The input voltage is usually a sawtooth waveform and the current output from the cell is fed into a current-to-voltage converter. By choosing a suitable value for $R_m$ the output voltage $V_o$ is made proportional to the current in the working electrode and can thus be displayed on an $X$-$Y$ recorder together with the input wave form $V_i$. The op-amp configured as a voltage follower is required to have a very high input impedance in order (ideally) to eliminate the current drawn from the reference electrode. It should be

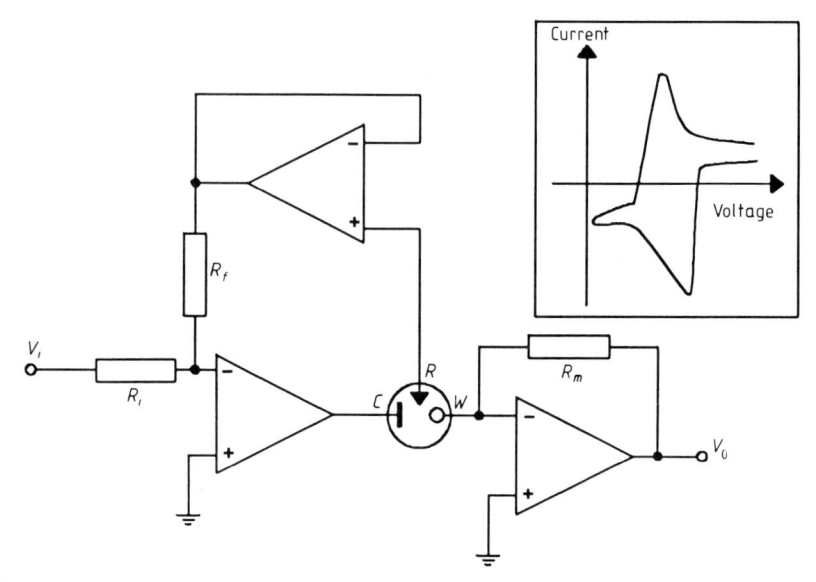

**Figure 3.29** A circuit diagram of a potentiostat and electrochemical cell. The inset shows a typical cyclic voltammogram obtained by plotting $V_o$ against $V_i$.

appreciated that the measured current from the electrochemical cell is very small (typically a few nanoamps or less) thereby necessitating the use of op-amps with very low bias currents. A plot of current versus voltage is referred to as a *polarogram* or *cyclic voltammogram* and to the trained user, the position of the peaks on the voltage axis indicate qualitative information whilst the magnitude of the current shows quantitative information about the solution under test.

### 3.3.3.5 *Humidity*

Humidity is a measure of the amount of water vapour present in a gas (usually air). Humidity sensors are required in a variety of different applications, particularly for the industrial apparatus used in the manufacturing of food and textiles. Relative humidity (RH) is the ratio of the actual water vapour pressure to the saturation vapour pressure at a certain temperature and is usually expressed in % RH. The dew point is the temperature at which the actual pressure and the saturation pressure are equal. Condensation would occur if the gas were to be cooled. The discipline of humidity measurement is called **hygrometry**.

Many humidity sensors are based on measurement of an electrical parameter such as resistance or capacitance of a material which changes when exposed to water vapour. Obviously the materials chosen must exhibit reproducible and reversible changes. An example of a simple device is shown in figure 3.26. Porous metal-oxide ceramic materials such as aluminium oxide have proved to be popular for the construction of capacitive humidity sensors.

The sensing elements of psychrometric sensors or wet-and-dry bulb devices, are temperature sensors, (platinum wire or thermistor) one of which is exposed to the ambient (dry bulb) and the other is covered by a wick which is saturated with distilled water (wet bulb). When air is passed over the wet bulb the sensing element is cooled below ambient temperature due to evaporation of water from the wick. The evaporation, hence temperature difference depends of the moisture content of the air. The RH can be determined from a psychrometric chart which is a plot of wet and dry bulb temperatures at a certain barometric pressure.

A popular way to determine absolute humidity is by measuring the dew-point temperature. A common approach is to cool a surface whose temperature is being monitored, until dew begins to condense upon it. Several methods exist to detect the presence of the dew, one way is to use a photoelectric method whereby the condensation surface is a polished mirror and a light beam is aimed at the mirror. When condensation occurs the light is scattered in different directions and this is detected by a number of photodetectors. Alternatively if the condensation surface has a metal electrode pattern deposited onto it, the presence of water can be detected by monitoring the capacitance change (water has a dielectric constant of around 80). Both these techniques require a suitable temperature sensor to be integrated into the condensation plate to measure the surface temperature. Cooling us usually achieved by using a Peltier device mounted underneath the condensation surface.

### 3.3.3.6 *Biosensors*

Biosensors have attracted a great deal of research activity. Technological breakthroughs in the area of micromachined sensors and actuators are expected to find greatest application for biomedical purposes. For example, a biosensor for measuring blood glucose level could be used in conjunction with a micromachined pump which would enable the correct dosage of insulin to be automatically administered to the patient. The complete system would be extremely small and portable. The number of commercially available solid state biosensors has been limited, but some are based on modified ISFET structures. An enzyme-selective membrane could be used for sensing, say urease. Another example of a biosensor is a micromachined silicon electrode sensor for the detection of neural activity in bees' brains (Pickard *et al* 1990). This device is a 12-channel probe micromachined in silicon to dimensions 800 μm by 150 μm wide by 20 μm in thickness which can be inserted into the brain of a honey bee and allow *in vivo* neuronal activity to be monitored.

Biosensors will continue to be the subject of intensive research activity for the foreseeable future, the challenges which these sensors present to the sensor designer can be quite severe, not only for the development of the sensing element but also in the design of the encapsulation and packaging methodologies.

## 3.3.4 Magnetic

Most magnetic field sensors use an effect first observed by Edwin Hall in 1879. He found that when a transverse magnetic field is applied to a current-carrying conductor an electric field results which is perpendicular to the current and also to the magnetic field. The resulting Hall voltage arises due to the Lorentz force acting on a charge carrier travelling in a magnetic field. The Hall effect in metals is much smaller than that associated with semiconductor materials like silicon and gallium arsenide.

Another phenomenon which is exploited for magnetic sensors is the magnetoresistive effect which can be observed in semiconductors, metals and ferro magnetic materials. When a magnetic field is applied across a magnetoresistor a corresponding change in resistivity (resistance) can be measured. Recent technological advances in the area of thin-film microelectronics have allowed magnetoresistive sensors to be developed that are suited for the mass market in view of their low-cost high sensitivity and low power consumption aspects.

### 3.3.4.1 *Hall effect*

If a current $I_x$ passes through a small thin slab of a metal or semiconductor with dimensions length $l$, width $w$ and thickness $t$ and a magnetic flux density $B_z$ is applied perpendicular to the direction of current flow then the resulting Hall voltage, $V_H$, can be measured across the contacts as shown in figure 3.30. This voltage is found to be proportional to the current and the magnetic field and is given by

$$V_H = \frac{R_H I_x B_z}{t} \tag{3.49}$$

where $R_H$ is the so-called Hall coefficient. The Lorentz force associated with an electron, having charge $(-e)$, moving with average velocity $v$ in a magnetic field $\mathbf{B}$ is given by

$$\mathbf{F} = -e(\mathbf{v} \times \mathbf{B}) . \tag{3.50}$$

The effect of the Lorentz force is to deflect electrons in a direction orthogonal to both the current and the magnetic field (the y-direction in the diagram). The resulting electric field opposes the Lorentz force and eventually equilibrium exists and the current flows in the original direction. The time taken to reach this position is of the order of $10^{-4}$ s. The Hall field which compensates the Lorentz force is given by

$$E_H = +v_x B_z \tag{3.51}$$

PHYSICAL PRINCIPLES OF SENSING 79

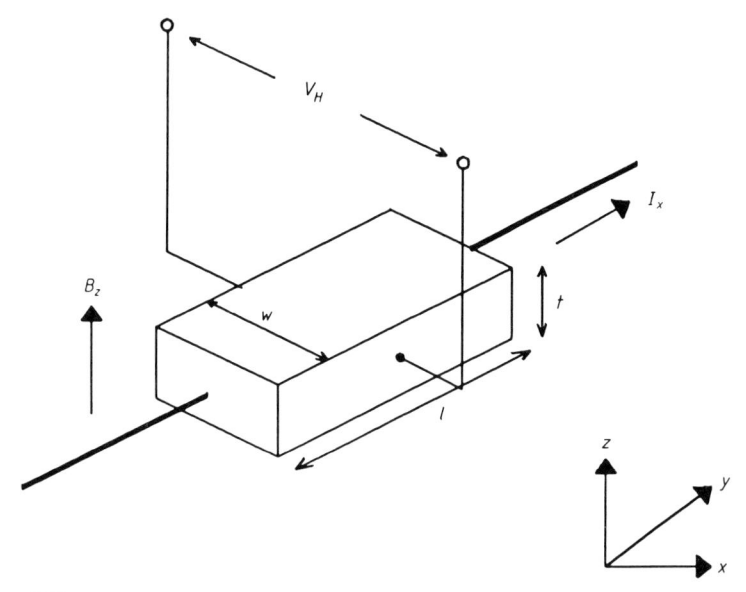

**Figure 3.30** Diagrammatic representation of the Hall effect.

The current density $J_x$ is

$$J_x = -nev_x \tag{3.52}$$

where $n$ is the electron density, hence the Hall field is

$$E_H = -\frac{J_x B_z}{ne} \tag{3.53}$$

which can be rewritten as

$$E_H w = -\frac{J_x wt B_z}{net} \tag{3.54}$$

thus giving

$$V_H = -\frac{I_x B_z}{net} \tag{3.55}$$

and the Hall coefficient is given by

$$R_H = -\frac{1}{ne}. \tag{3.56}$$

The Hall coefficient is inversely proportional to the carrier density which explains why the Hall effect is more pronounced in semiconductors than in metals. Indeed the minus sign in the above expression accounts for the charge carriers being electrons as in metals and $n$-type semiconductors. For $p$-type semiconductors the Hall coefficient is given by $R_H = 1/\rho q$, where $\rho$ is the hole density and $q$ the effective charge.

### 3.3.4.2 *Magnetoresistors*

Most metals and semiconductors exhibit a slight change in resistance when subjected to an external magnetic field. The effect is more pronounced with ferromagnetic materials and some semiconductors like InSb. The phenomenon is referred to as magnetoresistance. For classification purposes magnetoresistors are sub-divided into two categories namely, geometric and physical devices. With geometric magnetoresistors the resistance change is caused by geometrical factors like the length, width thickness and shape of the slice of material. Magnetoresistive elements which rely on physical process, like velocity distribution of charge carriers, are referred to as physical magnetoresistors. The latter effect, however, is rarely used in present day device applications.

If a rectangular magnetoresistive platelet is made such that the length $l$, is much smaller than the width $w$, and this is subjected to similar conditions to those in figure 3.30, then the end electrodes will tend to short-circuit the Hall voltage and so the Hall field will not balance the Lorentz force. Hence the charge carriers will tend to deviate from the $x$-direction thereby increasing the effective path length and also the resistance. In order to produce a magnetoresistor with a high sensitivity ($\Delta R/R$) a geometry needs to be adopted whereby the Hall voltage is short-circuited. Two common configurations are the Corbino disk and the barber pole arrangement depicted in figure 3.31.

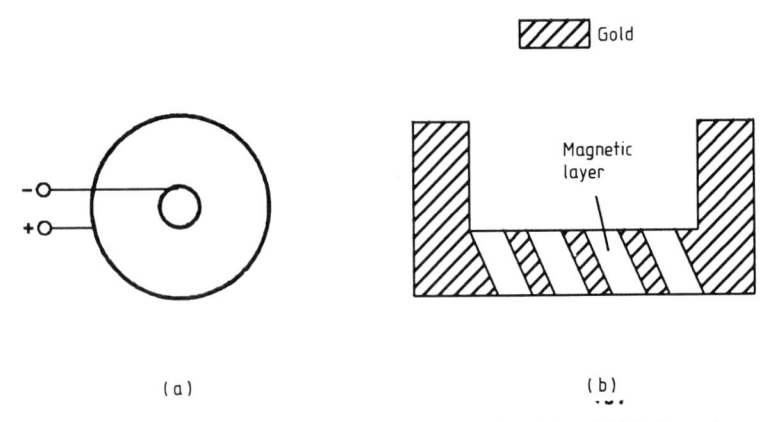

**Figure 3.31** Two configurations for magnetoresistors: (a) Corbino disk and (b) barber pole arrangement.

## PHYSICAL PRINCIPLES OF SENSING

The coaxial electrodes on the Corbino disk cause the $E_y$ field to be zero thereby eliminating the Hall voltage. The barber pole arrangement is effectively a series connection of magnetoresistive elements with a small length to width ratio. The elements are inclined in order of linearize the response to an applied magnetic field. Barber poles structures comprise a thin layer of magnetic material covered with stripes of gold electrodes. The magnetization in the thin films, in the absence of an applied magnetic field is said to lie in an *easy-axis*. When an external field is applied, the magnetization will rotate in the plane of the layer. When current flows through the film, the effect of an external magnetic field will be to alter the angle between the current and magnetization of the layer thereby resulting in a change of resistance of the film. The optimum angle between current and magnetization is 45°.

### 3.3.4.3 *Magnetotransistors*

The influence of a magnetic field on the various characteristics of many transistor structures give rise to a family of magnetic sensors known as magnetotransistors. Both bipolar and MOS technologies have been used to fabricate such devices. There is a distinction between devices in which lateral currents along the surface are influenced by a magnetic field and those in which the vertical component of current is more important. The general terms **lateral** and **vertical** magnetoresistors are thus used. In general bipolar devices utilize both there effects whereas with MOS transistors it is the lateral current which is predominantly used. Most transistor-based magnetic field sensors comprise a single emitter (or source) and several collectors (or drains) to sense the current. An example of a split-drain MOSFET Hall effect sensor is given by Cooper and Brignell (1984, 1985a,b). Here the authors describe an integrated magnetic field sensor fabricated using NMOS technology. A schematic representation of the current-splitting Hall sensor is given in figure 3.32. On application of an external magnetic field the Lorentz force causes electrons to be deflected towards one of the two drains thereby giving rise to two currents $I_1$ and $I_2$. The two drains were connected to two loads designed to give a shallow load line thus maximizing the voltage difference for a given current difference. Other types of magnetotransistors are covered in more detail elsewhere (Middlehoek and Audet 1989).

### 3.3.4.4 *Applications*

Magnetic sensors find application in many other areas as well as their direct use in the measurement of the magnitude and direction of magnetic fields. Non-contact methods of electric current measurement are often desirable and by using a suitable soft magnetic core and a magnetic field sensor like a Hall device in the air gap it is possible to construct a current measuring device. If the current-carrying conductor is placed inside the magnetic core the magnitude of the current is proportional to the Hall voltage.

During the 1980s the storage capacity of disk memories has increased dramatically. A bit density of around $10^6$ bits/cm$^2$ is now easily achievable and the

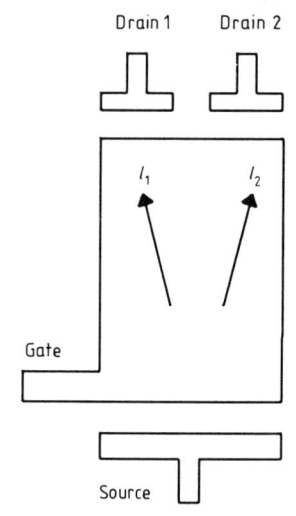

**Figure 3.32** Schematic structure of current splitting Hall sensor.

technology associated with the magnetic recording head has played an important part in achieving successful read/write activity. Many heads use thin films of magnetoresistive material. Such films also find application in reading the magnetic tracks on credit cards.

Hall effect devices produce a voltage which is proportional to the product of current and magnetic field. This multiplicative physical effect is rarely encountered and this peculiar phenomenon has been exploited in the construction of novel analogue multipliers.

### 3.3.5 Radiant

In general when one talks about the radiant signal domain it is electromagnetic radiation which is assumed to be the interest. Strictly speaking in this section nuclear particle radiation ought to be included, but in view of the limited space the discussion will be limited to the former type of radiant signal. Electromagnetic radiation includes radio waves, microwaves, cosmic rays, gamma rays, X-rays as well as UV, IR and visible radiation (light). The electromagnetic spectrum is depicted in figure 3.33.

The visible spectrum spans the region between 380 nm and 780 nm and is essentially the region to which the human eye is responsive. The band of wavelengths between 10 and 380 nm is termed ultraviolet radiation (UV) and the band between 780 and 10 nm infrared (IR) radiation. The IR band is sub-divided into *near* IR (780 to 3000 nm) and *far* IR (3000 nm to 10 nm). It is usual to show the spectral characteristics of electromagnetic radiation in terms of wavelength but

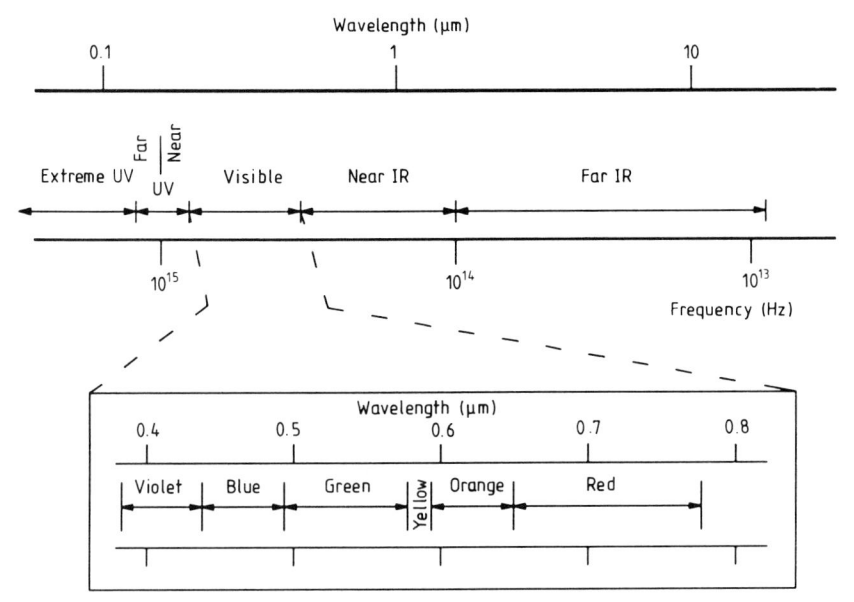

**Figure 3.33** The light spectrum.

other alternatives are in terms of frequency or photon energy ($E_p$). Photon energy is related to the frequency by Plank's constant $h$,

$$E_p = h\nu = \frac{hc}{\lambda} \tag{3.57}$$

where $E_p$ is in Joules, $\nu$ is in Hz, $h = 6.626 \times 10^{-34}$ Js, $c = 2.998 \times 10^8$ ms$^{-1}$ and $\lambda$ the electromagnetic wavelength (m). The measurement of electromagnetic energy is called radiometry and relates to measurement taken anywhere in the electromagnetic spectrum. Photometry deals with the measurement of electromagnetic energy only in the visible part of the spectrum allowance is made for the visible effectiveness of the light to the sensitivity response of the human eye (photo-optic eye response).

### 3.3.5.1 *Photovoltaic sensors*

This type of radiant sensor requires no external energy source and is based on the photovoltaic effect which concerns the generation of a voltage by incident light upon a junction between two dissimilar semiconductors. (Photodiodes and phototransistors will be discussed later). The junction acts as a potential barrier across which the flow of electrons is stimulated by incident photons. Material combinations such as selenium oxide/cadmium oxide and copper/copper oxide are often used in the construction of such devices. Figure 3.34 (a) gives a schematic

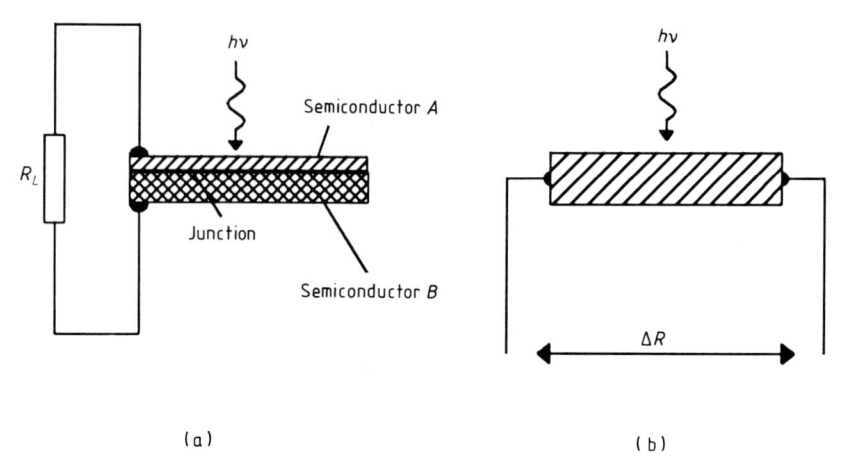

Figure 3.34 Examples of light sensors (a) photovoltaic and (b) photoconductive.

representation of a photovoltaic sensor. Semiconductor A is essentially transparent. The output voltage is taken across the load resistor $R_L$.

### 3.3.5.2 *Photoconductive sensors*
A material whose resistance changes in accordance with the amount of light incident upon it is referred to as photoconductive. Semiconductive materials like cadmium sulphide and cadmium selenide exhibit this property and are often used as light detectors. Increasing illumination results in a decrease in resistance which is usually measured across two electrodes on either side of the photoconductor. The change in conductance arises from a change in the absorption of the energy of incident photons figure 3.34 (b) shows a schematic representation of a photoconductor.

At any particular instant the absolute resistance of the material depends on a number of influencing factors including thickness, surface area, operating temperature, previous exposure levels and material and electrode geometry. Cadmium sulphide exhibits a non-linear resistance versus illumination characteristic and the non-linearity becomes more pronounced at higher illumination levels.

### 3.3.5.3 *Photoemissive sensors*
Photoemissive sensors emit electrons from a cathode as a result of photon absorption. Electrons are emitted from the surface when the photon energy is greater than the work function of the cathode material. Two devices which utilize the effect are the vacuum photodiode and the photomultiplier tube. In the vacuum photodiode the electrons are collected at an anode which is at a positive potential with respect to the cathode, a current thus flows and this can be measured by observing the voltage across an external series load resistor. Often further amplification is required as the direct output is quite small. The photomultiplier tube comprises

additional electrodes (dynodes) between the cathode and anode which are maintained at successively higher potentials with respect to the cathode. As a result of secondary emissions from the dynodes the current is amplified and can be measured by the ignition of and external resistor. Current gains of around $10^6$ are achievable and this is affected by the number of stages and the magnitude of the accelerating voltage.

### 3.3.5.4 *Pyroelectric sensors*
This type of radiant sensor consists of a ferroelectric material sandwiched between two electrodes. The material exhibits a spontaneous polarization which is temperature dependent. Incident radiant flux absorbed by the pyroelectric material results in a change of the material temperature, thereby producing a voltage across the electrodes which eventually decays to zero due to current flow through the internal leakage resistance. Most pyroelectric sensors are capacitive in nature and require the incident radiant flux to be chopped or pulsed. Materials such as lithium tantalite, triglycine sulphate and more recently polyvinylidene fluoride (PVDF) are often used as the pyroelectric material.

### 3.3.5.5 *Bolometric sensors*
These devices, like pyroelectric sensors, primarily use temperature variations to sense the radiant flux. A matched pair of thermistors is commonly used in a bridge configuration to measure the temperature. One of the thermistors is blackened and mounted such that it is exposed to the radiant signal. The other thermistor is isolated from the radiant flux and exposed to a reference temperature. The output voltage of the bridge is thus a function of incident radiant flux.

### 3.3.5.6 *Photodiodes and phototransistors*
The electrical characteristics of most diodes and transistors are affected by light. This is one reason why they are encased in a suitable opaque package. Phototransistors and photodiodes are constructed to enhance this property. Taking the example of a semiconductor *p-n* junction, a high electric field will exist across the depletion layer. When photons are absorbed on the surface to produce carrier pairs, the electric field at the junction will direct the holes towards the *p*-type material and the electrons will flow toward the *n*-type material. The resulting charge carrier unbalance causes a potential difference to occur between the two surfaces. This is the photovoltaic mode of operation. Alternatively, the device can be operated under reverse bias conditions thereby entering the photoconductive mode of operation. The choice of operating mode depends on system requirements. In general, the photovoltaic mode is chosen when low-level light detection is required. The photoconductive mode provides a faster response with a large linear dynamic range and better stability. Phototransistors provide amplification of the photon-induced current, but have a slower response time than a photodiode. A majority of photo-

diodes are made from silicon which gives a spectral response in the near IR region at around 800 nm.

PIN diodes contain an intrinsic layer sandwiched between the *p*- and *n*-type regions. The purpose of the intrinsic layer is to act as a sensitive volume for the absorption of photons. A sufficiently high reverse bias voltage needs to be applied to the device to ensure that the depletion layer extends throughout the thickness of the intrinsic layer. A high electric field across the intrinsic layer allows shorter collection times of the photogenerated carriers. PIN diodes also have a lower junction capacitance than a normal *p-n* junction device due to the depletion region depth.

Since the late 1960s, there has been considerable interest and advancement in the production of one-and-two-dimensional arrays of the photodiodes for applications in surveillance optical recognition, flow detection and aerospace guidance systems. Advances in VLSI technology have allowed the fabrication of arrays of photodiodes together with the associated electronic circuitry necessary for sampling and controlling each element within the array. The major drawback of such devices is their poor sensitivity. Other problems associated with parasitic capacitances limit the sampling rate.

### 3.3.5.7 *Charge-coupled devices (CCDs)*
A CCD is essentially a shift register formed by a series of closely spaced MOS capacitors. The CCD can store and transfer charge which has been introduced electrically or optically. The basic structure is depicted in figure 3.35. A series of electrodes, or gates, is fabricated on an insulator, typically $SiO_2$, which has been formed on a semiconductor substrate (such as Si). The gates are periodically interconnected. On the application of a phase-shifted clock, it is possible to shift the stored or generated charge along the silicon. A three-phase clock is required to ensure unidirectional travel and by altering the clock phasing it is possible to control the charge transfer direction. One photosensitive element is thus comprised of three MOS capacitors. The fraction of total charge remaining per transfer is called the transfer inefficiency. Another important performance feature is the maximum transfer rate. A number of different CCD structures which have been developed to improve factors like transfer efficiency and maximum transfer rate, these are the surface-channel CCD (SCCD), the buried-channel CCD, (BCCD) and the peristitial CCD (PCCD).

CCDs offer a low cost high resolution approach to image sensing. Many modern day video camcorders now use CCD arrays as the image sensors. The CCD elements used for this application exhibit high charge transfer efficiencies and transfer rate, low thermal noise and a flat spectral response in the visible spectrum.

*PHYSICAL PRINCIPLES OF SENSING* 87

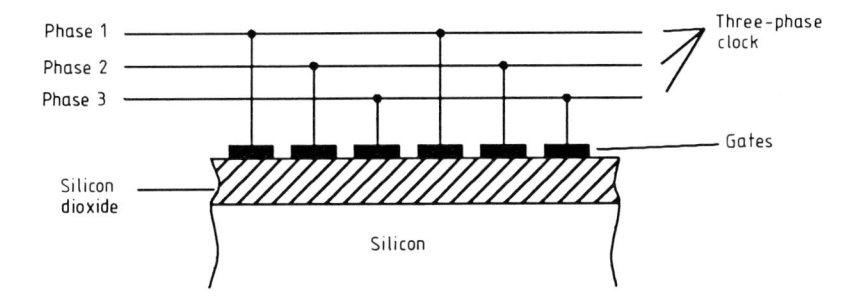

**Figure 3.35** Schematic of a three-phase CCD implemented in MOS technology.

# 4

# Electronic Measurement Techniques

## 4.1 TRANSDUCER INTERFACE CIRCUITS

In this chapter we shall discuss methods for facilitating measurement by electronic techniques. Many of the examples described in this chapter are well established methods for implementing the essential processes in traditional sensor systems - data acquisition, signal conditioning, amplification etc, and these fundamental concepts are equally important to intelligent sensor systems. Once again, because this area is more than adequately covered by existing texts, we shall present a condensed appraisal of a selection of the accepted techniques for electronic measurement, but in the light of the requirements of intelligent systems. The main aim therefore, is to outline the important procedures and guide the reader to sources of additional, more comprehensive coverage of each topic.

### 4.1.1 Bridge circuits

The Wheatstone bridge is the basis of many classical techniques of measurement and older measurement books are replete with variations on a theme. The great attraction of the bridge topology is that it embodies the principle of structural compensation by design symmetry (§ 3.2). Figure 4.1 shows a bridge circuit with four resistive elements and a voltage supply $V_s$. The electrical equivalent circuit is also shown.

An expression for the output of a generalized unbalanced resistive bridge with no load on the output can be obtained by reference to the Thevenin equivalent circuit. The values in the equivalent circuit are

$$E_T = V_s \left\{ \frac{R_1}{R_1 + R_4} - \frac{R_2}{R_2 + R_3} \right\} \qquad R_T = \frac{R_2 R_3}{R_2 + R_3} + \frac{R_1 R_4}{R_1 + R_4} \qquad (4.1)$$

and the well known relationship between resistances in a balanced bridge is

ELECTRONIC MEASUREMENT TECHNIQUES 89

$$\frac{R_4}{R_1} = \frac{R_3}{R_2} \quad (4.2)$$

Let us suppose that we have a transducer comprising four resistive elements of which only one responds to the measurand, while the others are fixed. We shall also assume that each element has a nominal resistance $R$. If the active resistor is inserted in place of $R_1$ in figure 4.1 and it changes its resistance by an amount $\Delta R$ in response to the input, then the output of the bridge will be

$$E_T = \left\{\frac{\Delta R}{4R + 2\Delta R}\right\} V_s \quad (4.3)$$

where $V_s$ is the bridge excitation voltage. It should be noted that this equation is non-linear in $\Delta R$, though it is usual to assume that $\Delta R$ is much less than $R$ so that the output of the quarter bridge is given by

$$E_T = \frac{\Delta R}{4R} V_s \quad (4.4)$$

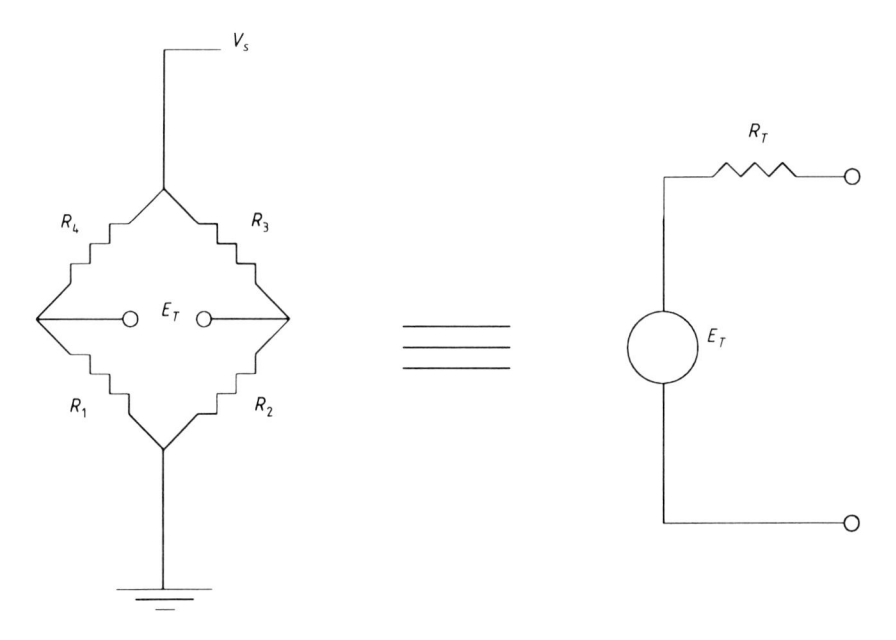

**Figure 4.1** The classic Wheatstone bridge and its electrical equivalent circuit.

and in general terms we have

$$E_T = \frac{n\Delta R}{4R} V_s \qquad (4.5)$$

where $n$ is the number of active arms. For a full bridge with four active arms we can get four times the signal and also four times the signal/noise ratio that can be obtained from one active arm. Also the equation is exact and linear in this case. We note at this stage that a common way of balancing a bridge is by the the addition of an extra resistor in parallel with one of the active arms. The consequence of this is that the response of the bridge again becomes non-linear.

An obvious disadvantage of the bridge topology is that it gives a balanced rather than a single ended output. Normally an operational amplifier is used to amplify the bridge output. It is advantageous to use one in conjunction with a tracking dual voltage generator (i.e. one which applies equal voltages to opposite ends of the bridge (Shepherd 1981)). The total sensor sub-system is shown in figure 4.2.

The selection of the impedance level at which to operate a bridge is a trade-off; provided that the impedance of the sensing element is disposable. In the example of strain gauges, the foil type allows very little variation of resistance, whereas the thick-film type allows variation over decades. High resistance is indicated where power consumption is an important consideration, but low resistance offers better immunity from thermal noise and pick-up. When the requirement for low power is combined with an enforced selection of a low resistance element the solution is to excite the bridge with a pulse train of low mark/space ratio. This technique can be combined with coherent detection, i.e. a gated amplifier controlled by the excitation train, to improve noise performance.

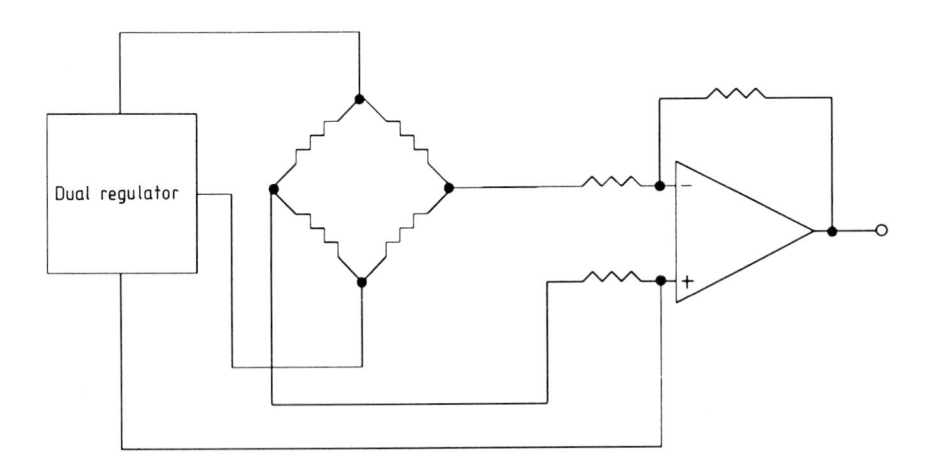

**Figure 4.2** An example of a method for generating a single ended output from a bridge.

# ELECTRONIC MEASUREMENT TECHNIQUES

An important variation of the Wheatstone bridge is the capacitance bridge, as a significant class of primary sensors convert variations of the target variable to variations of the parameter of capacitance, e.g. humidity (§ 3.3.3.5). The equation for the output voltage of a capacitance bridge is the same as equation (4.5) above, with $R$ replaced by $C$. There are two differences of technique which have to be borne in mind with capacitance bridges. Firstly, the excitation has to be an alternating voltage, ideally but not necessarily sinusoidal. Secondly, stray parameters are much more important, since the sensor capacitance and its variations tend to be small. Thus the electronic circuits associated with capacitance bridges are generally required to be physically close to the primary sensor.

## 4.1.2 Non-linear bridge elements

If the bridge circuit in figure 4.1 comprises a thermistor $R_1$, and dummy resistors $R_2, R_3$ and $R_4$, then by careful design it is possible to obtain a linear response from this bridge. Although we have versatile methods of linearization in sensors by digital processing, a first pass at linearization in the analogue section can improve the utilization of the precision of data conversion (§ 3.2).

If the thermistor exhibits a change in resistance $\Delta R$, in response to a temperature change $\Delta T$, then the output of the bridge is given by

$$E_T = V_s \left\{ \frac{R_1 + \Delta R}{R_1 + \Delta R + R_4} - \frac{R_2}{R_2 + R_3} \right\}. \tag{4.6}$$

If the bridge is balanced at a reference temperature then

$$R_4 = \frac{R_3 R_1}{R_2}. \tag{4.7}$$

Therefore,

$$E_T = V_s \left\{ \frac{1 + \frac{\Delta R}{R_1}}{1 + \frac{\Delta R}{R_1} + \frac{R_3}{R_2}} - \frac{1}{1 + \frac{R_3}{R_2}} \right\} \tag{4.8}$$

which can be rewritten as

$$\frac{E_T}{V_s} = \left\{ \frac{1 + \gamma}{1 + \gamma + r} - \frac{1}{1 + r} \right\} \tag{4.9}$$

where $\gamma = \Delta R/R_1$ and $r = R_3/R_2$.

This clearly exhibits a non-linear response, the degree of which can be adjusted by selecting values of $\gamma$ and $r$. However, $\gamma$ is usually well defined as it is related to the material properties of the thermistor. Therefore, by choosing a suitable ratio $R_3{:}R_2$ and selecting an appropriate supply voltage $V_s$, the response can be tailored to approximate to a linear characteristic.

### 4.1.3. Low-power interfacing

Occasionally there is a requirement for minimizing the electrical power consumed by a sensor and its interface electronics. A prominent example is the case of intrinsically safe applications such as mining. Optical fibres a clearly offer advantages in terms of electrical noise immunity, but they also allow the transmission of both signal and power without the use of metal conductors, thereby eliminating sparking hazards. For these reasons hybrid optical fibre sensors have proved to be a popular form of low-power sensor. Hybrid, as opposed to intrinsic, devices use conventional electrical sensors such as strain gauges, platinum resistance thermometers etc, together with optical fibres for both the data transmission and for carrying the power to the sensor and its interface. Figure 4.3 shows an outline schematic diagram of an optically powered hybrid optical fibre sensor devised by Ross (1991). If the optical power is provided by a single photodiode, the voltage

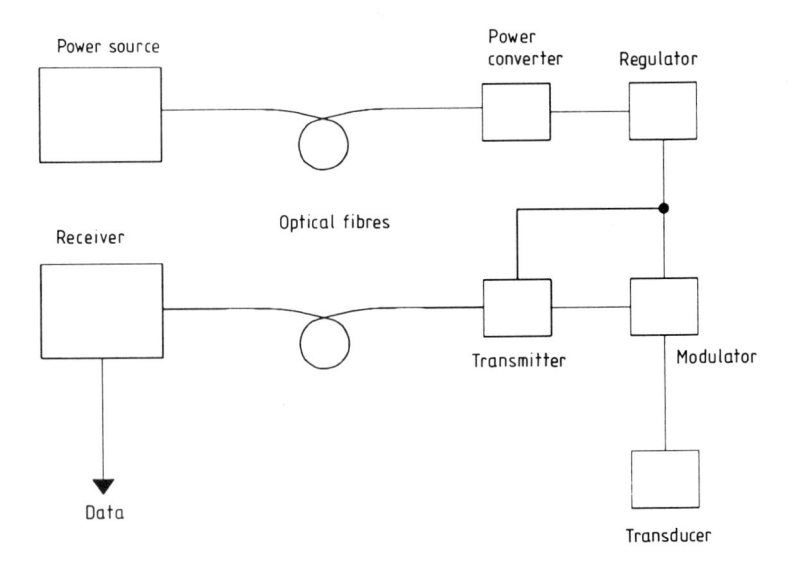

**Figure 4.3** An optically powered hybrid optical fibre sensor (after Ross 1991).

ELECTRONIC MEASUREMENT TECHNIQUES

will not normally be high enough for an active interface circuit. It is therefore necessary to increase the voltage to a suitable value.

The optical source supplies power to the sensor head via the optical fibre. After conversion to electrical power it is then regulated to supply excitation to the transducer and its interface. The silicon photodiode is often used for the power conversion, and such devices produce a terminal voltage of around 0.5 V. This voltage can be increased by adopting stepping up techniques similar to those employed in switch mode power supply design such as forward and flyback conversion. By such techniques it is possible to produce circuits which have an input optical power of 2 mW or less, and yet can supply enough voltage to run, say, a resistance thermometer and its interface circuit.

## 4.2 OPERATIONAL AMPLIFIERS

A key component in the analogue section of the smart sensor is the operational amplifier (op-amp). This versatile sub-system can be used in a variety of configurations, some of which are discussed below. The ideal op-amp is a device with infinite gain, zero output impedance and infinite input impedance; and, while it is tempting to assume that these conditions apply, it is important to bear in mind the

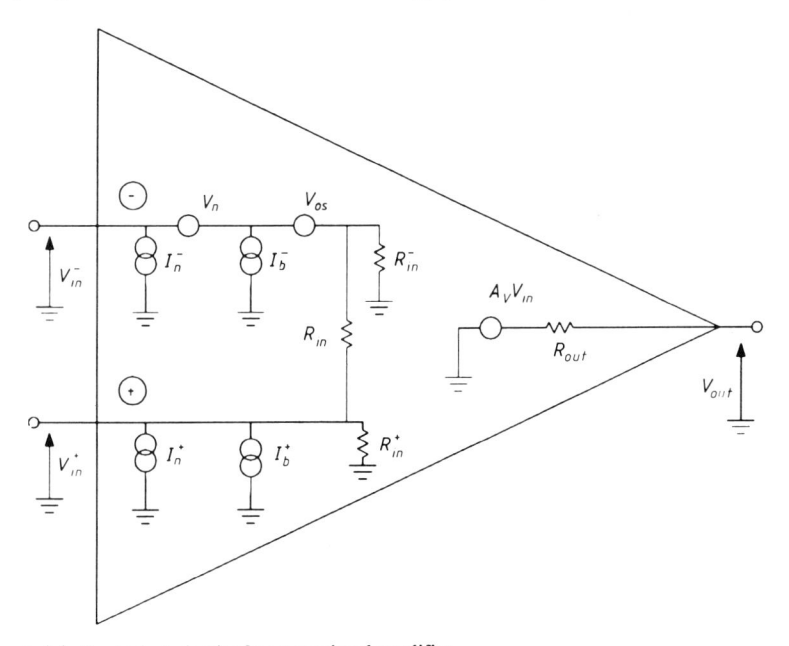

**Figure 4.4** Equivalent circuit of an operational amplifier.

defects in the behaviour of all op-amps. These may be summarized in the form of an equivalent circuit as shown in figure 4.4.

It is not appropriate to deal with the detail of this circuit here, and a fuller discussion can be found in the literature ( Shepherd 1981,Watson 1989). Suffice it to note that with each input there is an associated noise source, a bias current source and a common mode input resistance; while differentially there is a voltage noise source, an offset voltage and differential input resistance. There is also a non-zero output resistance.

Each of these elements can become important in sensing systems under the appropriate conditions, and they have to be carefully considered in the trade-offs associated with the analogue sub-system design process. The variety of analogue sub-systems that can be created with op-amps is quite considerable (Shepherd 1981, Horrowitz and Hill 1990); examples include active filters, oscillators, voltage regulators, sample-hold circuits, integrators etc. The following sections present some of the more important applications of op-amps to intelligent sensor systems.

### 4.2.1 Instrumentation amplifier

This configuration is a key element in many instrumentation systems. It is usually used in the differential mode of operation, e.g. to amplify the output from a bridge circuit. Amplifiers of this nature tend to have high input impedances and high common mode rejection ratios (CMRR), i.e. they have a high differential gain but tend to reject common mode signals like noise, bridge offset voltage etc. A common arrangement is the so-called cross-coupled amplifier which is shown in figure 4.5.
Analysis of this circuit reveals that the differential output of the input stage is given by

$$V_1' - V_2' = \left(1 + \frac{2R_2}{R_1}\right)(V_1 - V_2). \tag{4.10}$$

We can see that for common mode signals (i.e. $V_1 = V_2$) we have $V_1' = V_1$ and $V_2' = V_2$, indicating that common mode signals are passed at unity gain. The single ended output voltage is thus

$$V_o = n\left(1 + \frac{2R_2}{R_1}\right)(V_1 - V_2). \tag{4.11}$$

It is usual practice to have the highest gain at the front-end amplifiers, thereby making $n$ unity. The success of this circuit relies on the fact that there are good matching characteristics between the op-amps. A number of commercial integrated circuits are available which are based on the above circuit. although more elaborate forms are frequently used. Variations on the theme include the addition of capacitors in parallel with $R_2$ and $R_3$ to alter the bandwidth of the amplifier, and the

ELECTRONIC MEASUREMENT TECHNIQUES  95

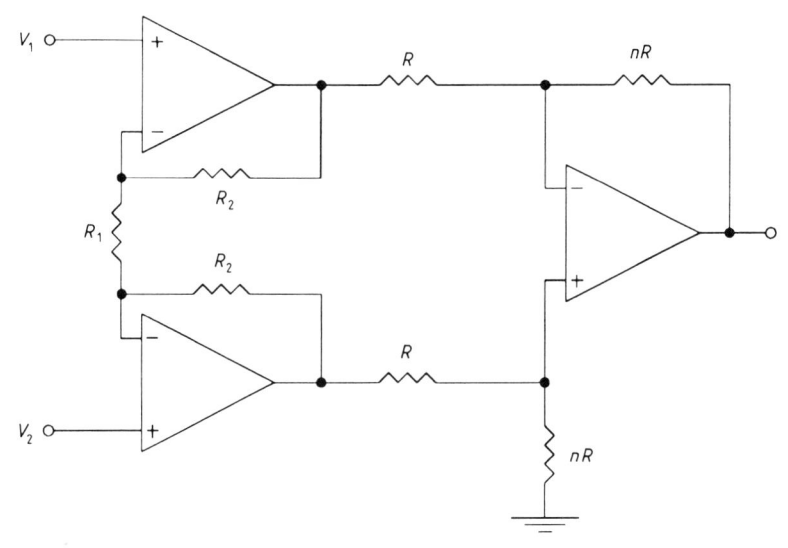

**Figure 4.5** The cross-coupled follower instrumentation amplifier.

inclusion of precision trimmed resistors which provide sub-systems with very accurate gain selection. Some high performance devices have additional circuitry to improve the stability and reduce parameter drift with respect to both time and temperature.

### 4.2.2 High performance amplifiers

For low drift applications, differential chopper stabilized and auto-zeroing op-amps are available. The differential chopper input amplifiers use internal sampling and self zeroing techniques to deal with the error voltages, and sample-and-hold circuits to ensure signal continuity during the error correction intervals. Offset voltage drifts of around 0.3 µV/°C and 1 µV per month are typical for this type of amplifier. They are well suited to bridge circuit applications.

Auto-zeroing amplifiers exhibit a similar performance specification to chopper stabilize op-amps but they offer the advantage of having the offset storage capacitors on-chip. Figure 4.6 gives a block diagram of such an amplifier and illustrates how the auto-zeroing is achieved. The four main components are: the main amplifier, output buffer, offset correction amplifier and internal oscillator in addition to the on-chip capacitors $C_1$ and $C_2$. The pin-out of theses amplifiers tends to be compatible with general purpose op-amps such as the 741. The internal oscillator commutates the switching signals to the switches A and B. During one phase, switch A closes to connect the amplifier input voltage to $C_1$, which charges up until the voltage across it equals the offset voltage. During the next phase, switch A is disconnected from the input and switch B closes to transfer the charge stored on

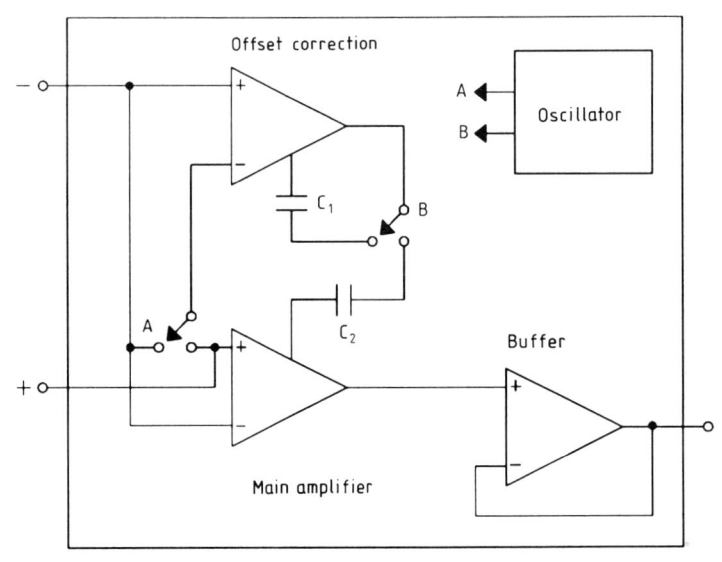

**Figure 4.6** Block diagram of an auto-zeroing operational amplifier.

$C_1$ onto $C_2$. The voltage across $C_2$ is subtracted from the output of the main amplifier thereby compensating for any long-term drift in the offset voltage. Typical devices exhibit a 5 µV offset voltage and a drift of 0.05 µV/ °C. A bandwidth of 1.5 MHz at unity gain is not uncommon.

### 4.2.3 Isolation amplifier

Sometimes situations arise where there is a need for coupling an analogue signal from a sensor into another circuit with a different ground reference. One reason for taking such action might be where there are severe electrical disturbances on the sensor ground line. Another common need is in the area of medical electronics where the human subject must be isolated electrically from any instrument powered directly from the mains.

Isolation amplifiers are devices in which the input is effectively isolated from the output by a transformer or optical link. In rare instances, the potential difference between the input 'ground' and the output 'ground' may exceed several thousand volts. Figure 4.7 shows the basic concept of the isolation amplifier. If the isolation barrier is transformer-based then the signal must first be modulated at the input and subsequently de-modulated at the output. Opto-isolation is usually achieved by a LED at the sending end and a photodiode at the receiver.

ELECTRONIC MEASUREMENT TECHNIQUES

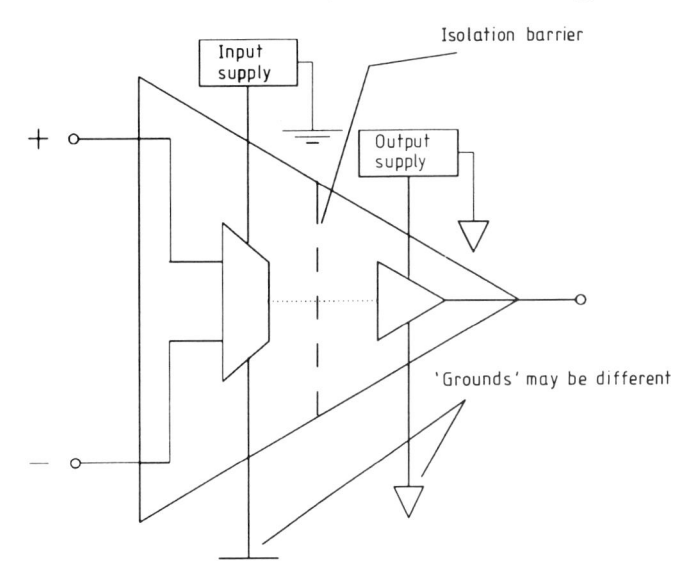

**Figure 4.7** Functional block diagram of an isolation amplifier.

### 4.2.4 Logarithmic amplifier

Sometimes there is a requirement to measure a physical quantity over a wide dynamic range, yet the information has to be squeezed through a restricted window, such as an ADC. One approach to this problem is gain switching, but an alternative is to use a continuous compression device. Logarithmic amplifiers are often used to achieve this aim. Figure 4.8 shows a schematic of a logarithmic amplifier implemented with an op-amp, a bipolar transistor and a resistor.

The basis of the circuit operation is that an exponential relationship exists between the transistor collector current $I_c$, and the base-emitter voltage $V_{BE}$. The following equation is valid over several decades.

$$I_c = I_{ss} \exp\left(\frac{qV_{BE}}{kT}\right) \tag{4.12}$$

where $I_{ss}$ is the reverse saturation current, $q$ is the electronic charge, $k$ is Boltzmann's constant and $T$ is the absolute temperature.

Assuming that the op-amp has a very high input impedance and that no current flows into its input terminals we have,

$$\frac{V_i}{R} + I_c = 0 . \tag{4.13}$$

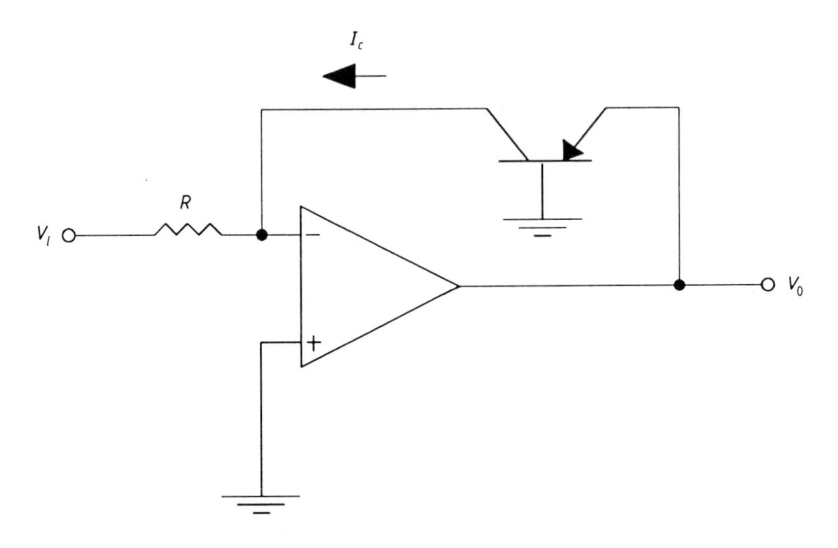

**Figure 4.8** A simple logarithmic amplifier.

By noting that $V_o = V_{BE}$ and substituting equation (4.12) into the above, the following relation is obtained,

$$V_o = -\frac{kT}{q} \log_e \left(\frac{V_i}{RI_{ss}}\right). \tag{4.14}$$

The output voltage is therefore proportional to the natural logarithm of the input voltage. Clearly, there is a strong temperature dependence on this relationship, and commercial devices compensate for this by by including an additional transistor in the circuit to provide a correction voltage. The two transistors are usually a matched pair which are thermally coupled.

By interchanging the transistor and the resistor it is possible to realize an antilogarithmic amplifier. Using a similar analytical argument to that shown above, the expression relating $V_o$ and $V_i$ for the antilog amplifier is given by,

$$V_0 = -RI_{ss} \exp\left(\frac{V_i}{kT}\right). \tag{4.15}$$

Once again, there is a large temperature dependence whose effect is minimized in a commercial product.

### 4.2.5 Charge amplifier

In this discussion so far, we have considered electronic measurement techniques for mainly resistive-based sensing elements. For high impedance, capacitive-type sensors the previous circuits are not always suitable. One particular example is a piezoelectric transducer which gives rise to a charge Q when subjected to an applied force. The charge at the electrodes results in a voltage $V_p = Q/C_p$, where $C_p$ is the capacitance of the piezoelectric element. Initially, it would appear that this voltage could be amplified by a conventional voltage amplifier. In practice however, $V_p$ is vulnerable to unwanted noise, pick-up etc, due to the capacitive nature of the piezoelectric material and the associated cabling $C_c$.

To overcome this problem a charge amplifier is often employed. Figure 4.9 shows a charge amplifier connected to a piezoelectric sensing element which is modelled as a current source $i_p$, in parallel with $C_p$. The capacitance of the cable is also shown. The feedback resistor $R_f$, which usually has a relatively high value ($> 10^8 \, \Omega$), provides a path for direct current. If this were omitted the capacitor $C_f$ would steadily charge, by virtue of the small bias current in the op-amp, until the amplifier saturated. The basic configuration is that of an op-amp integrator, hence $V_o$ is proportional to the charge developed across the plates of the piezoelectric material. The virtual earth at the op-amp input ensures that there is effectively no

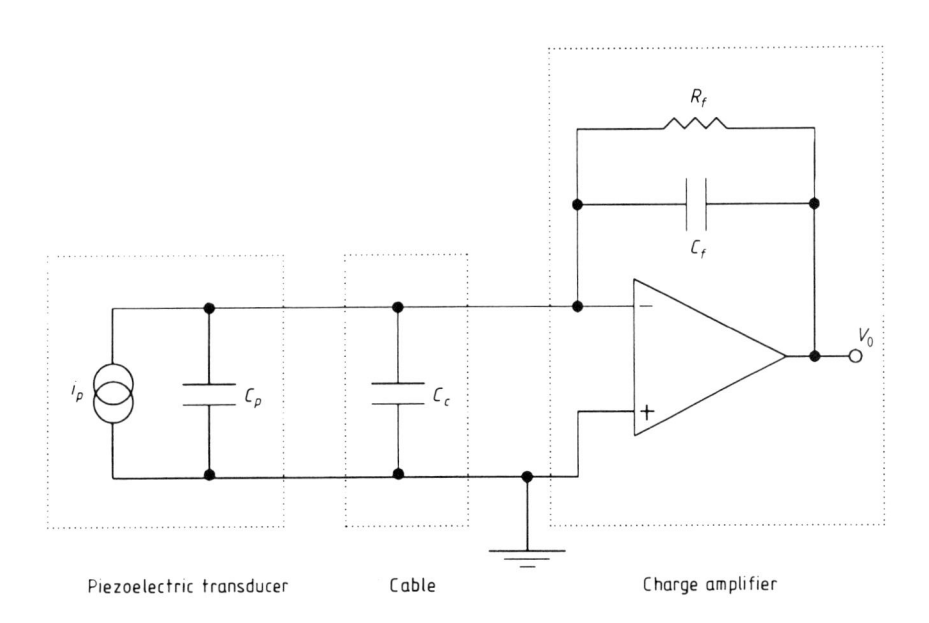

**Figure 4.9** A charge amplifier.

voltage drop across either $C_p$ or $C_c$, so the cable has little effect on the overall gain of the amplifier. Long cables may be used with charge amplifiers as the cable capacitance does not result in a reduced sensitivity or a variation in the frequency response.

## 4.3 DATA CONVERSION

We have adopted the definition that the intelligent sensor will contain a digital processor, usually in the form of a microprocessor. Interfacing between analogue and digital quantities, and vice versa, therefore becomes a necessity. In this section we will provide a short appraisal of the various types of analogue-to-digital converter (ADC) and digital-to-analogue converter (DAC). Although a full review of all possible techniques is beyond the scope of this book, we attempt to highlight some of the useful criteria which need to be considered in the selection of ADCs and DACs.

### 4.3.1 Digital-to-analogue converters (DACs)

We begin our discussion with DACs since they are used as components in many ADCs. A fundamental property of any DAC, or for that matter ADC, is the number of bits for which it is designed. This determines the resolution of the device and the information capacity of the system in which it is embedded. For a DAC the resolution is the smallest change that can occur in the analogue output as a result of a change in the binary input. The accuracy of a converter is essentially the difference between the actual output and the theoretical output, and is dependent upon the precision of the components within the device.

#### 4.3.1.1 *Binary weighted ladder*
Figure 4.10 shows a simple 3-bit DAC comprising an op-amp and several resistors. The three input resistors are connected together at the negative input of the op-amp forming a current-summing junction. Each input resistor is twice the value of the previous one, hence they are termed *binary weighted*. If a voltage $V$ is applied to each of the inputs in turn, the value of $V_o$ would be $-V$, $-V/2$, $-V/4$ respectively. In general terms, the output is proportional to the weighted sum of the input voltages. The source for the input voltages must be both stable and accurate. The lowest value resistor affects the most significant bit (MSB) and must therefore have the highest precision. In principle, the basic idea may be extended to increase the resolution of the device. In other words, the number of input resistors will be equal to the number of bits. However, for a high precision DAC, say 16-bits, this would require 16 high precision resistors ranging in values from $R$ to $65536R$. This is

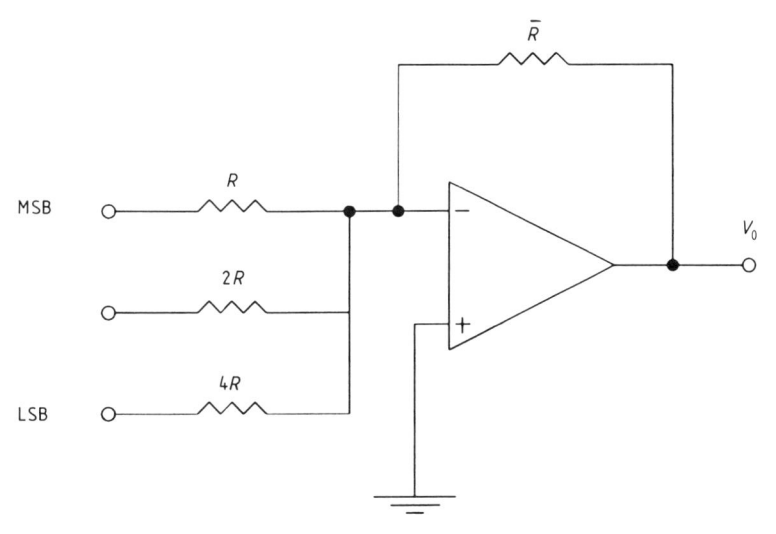

**Figure 4.10** A binary weighted ladder DAC.

extremely difficult to achieve in practice, and so the summing junction DAC is generally only used in lower precision applications.

### 4.3.1.2 *R-2R ladder*
A solution to the previous problem associated with the weighted ladder technique is to use an *R-2R* ladder. This configuration utilizes twice the number of resistors for the same precision, but is generally considered to be a more elegant solution. Furthermore, only two values of resistance are needed, so an accurate ratio of values is required but the absolute values are not as important. Figure 4.11 shows a typical *R-2R* ladder DAC.

The ladder is used as a current splitting circuit, and it has the property that the resistance at any node, with the others grounded, is *3R*. If the op-amp feedback resistor $R_F = R$, then the voltage gain of the circuit is -1. The electronic switches are used to apply either 0 volts or a reference voltage $V_{ref}$ to the inputs of the ladder. For example, if the binary word 0001 was applied to $I_0$, $I_1$, $I_2$, $I_3$ respectively, then using Thevenin's theorem it can be shown that the equivalent circuit of the ladder is a voltage source $V_{ref}/16$ in series with a *3R* resistor. This gives rise to an output voltage of $-V_{ref}/16$ volts. If the binary value 1000 was applied to the inputs then the analogue output voltage would be $-V_{ref}/2$ volts. The argument can be followed through to show that, in general, the output of the DAC is a voltage proportional to the binary input value.

This type of DAC is more amenable to high precision applications than the previous example, provided that a stable reference voltage is used. As an example, if the reference voltage was 8 volts and 12-bits were used, this would provide a resolution of around 0.02 %, which produces a 2 mV step size at the output. For

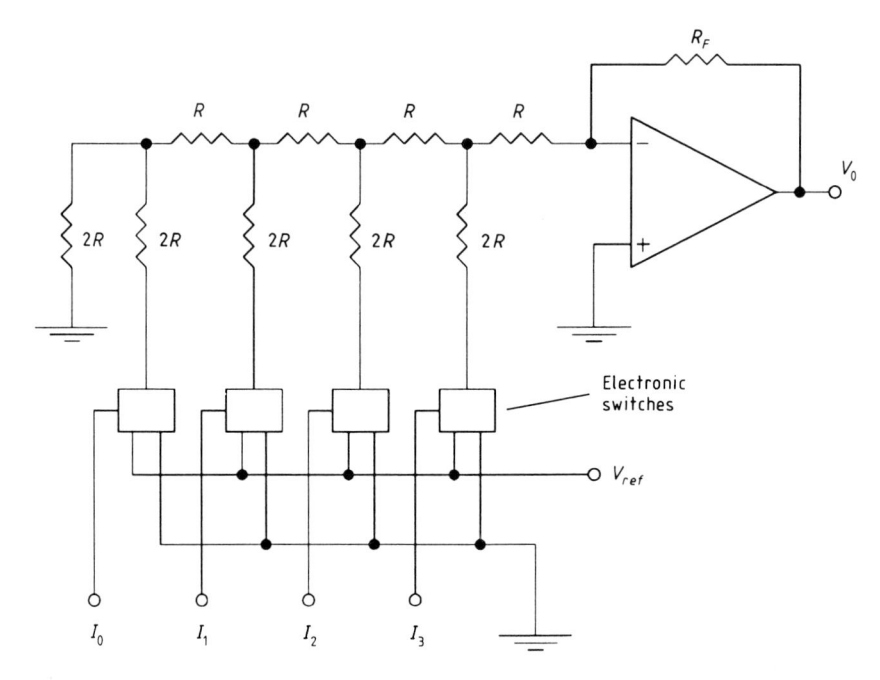

**Figure 4.11** The *R-2R* ladder DAC.

high frequency applications, the performance of the *R-2R* ladder is limited by the quality of the op-amp and the stray capacitance to ground.

### 4.3.2 Analogue-to-digital converters (ADCs)

We will now take a closer look at some of the fundamental principles behind ADCs. Although we would not normally expect to design such a converter from scratch, it is important to understand how ADCs operate in order to be able to make an appropriate selection for a particular application. Four distinct techniques will be covered in this section namely; ramp-type, successive approximation, dual slope integrating and flash converters.

#### 4.3.2.1 *Ramp-type ADCs*
This type of analogue-to-digital converter is sometimes referred to as a voltage-to-time converter. The three main elements are a binary counter, a DAC and an analogue comparator. A 4-bit example is depicted in figure 4.12. During the conversion time the analogue input voltage must not be allowed to change and use is made of a sample-and-hold (S/H) circuit which samples the input waveform

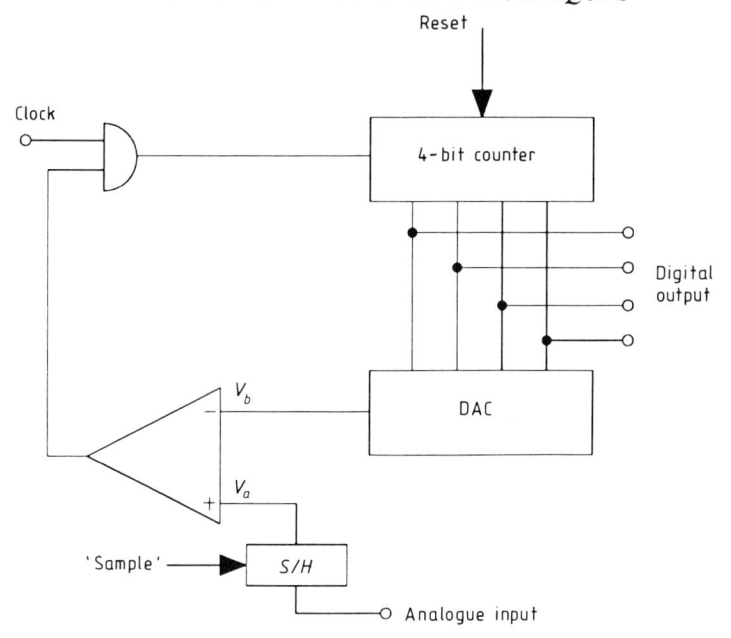

**Figure 4.12** Ramp-type ADC.

when a valid *sample* pulse is applied. The S/H circuit holds the analogue signal at its captured value during conversion.

Initially the counter is reset and the analogue input is sampled. The output of the DAC is thus zero, and $V_a > V_b$, so the output of the comparator is a logic 1. The AND gate is therefore enabled and the binary counter begins to count up. This process continues until the output of the DAC is greater than the analogue input voltage ($V_b > V_a$). At this point the output of the comparator is a logic 0 and hence the counter is inhibited. The equivalent binary value of $V_a$ is presented at the output.

This is an example of an early type of ADC and its main disadvantage is that it is relatively slow. For a device with a resolution of *n*-bits the conversion time could be up to $2^n$ clock cycles. In general, the conversion time varies with the value of $V_a$.

### 4.3.2.2 Successive approximation ADCs

This is one of the most popular type of ADC currently available. A schematic diagram is shown in figure 4.13. The main elements are a DAC, a successive approximation register, an analogue comparator and a control logic module. The successive approximation register supplies the DAC with various output codes and the result is compared with the analogue input via the comparator. Also included with this device is a start conversion input and an end conversion output.

Initially the successive approximation register outputs a value corresponding to half the maximum input value. This is achieved by setting the most significant bit

104        INTELLIGENT SENSOR SYSTEMS

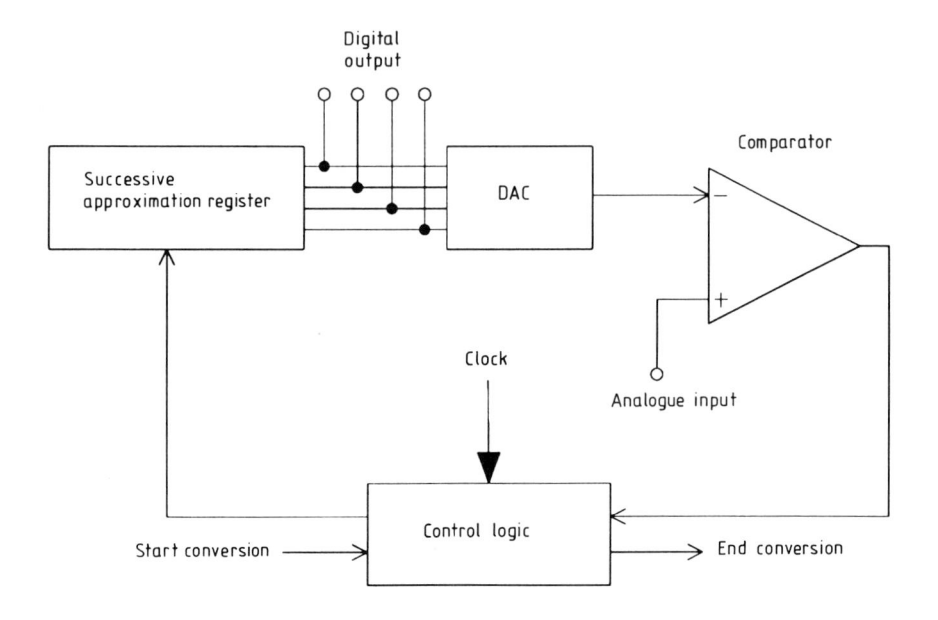

**Figure 4.13** Successive approximation ADC.

(MSB) to a logic 1 value. If the analogue input is greater than this value then the MSB remains at 1 and the next lesser significant bit is set to 1. However, if the input is less than the MSB value, the MSB is set to a logic 0 and the next smallest bit is set to a logic 1. If this value proves to be lower than the input voltage then the next smaller bit is set to a 1 and so the process continues until the binary output corresponds to the analogue input. Essentially a binary search is initiated on the receipt of a suitable start conversion pulse. Once the process is complete the end conversion line is set and the digital output word can be used.

Figure 4.14 shows a graph of the output values against number of clock pulses for a 4-bit example. Assume that an analogue voltage of 10.3 volts is applied to the device. If the maximum input was, say, 16 volts, then the initial guess would be 8 volts. The analogue input is greater than this so this bit is retained and the next smallest bit is set, producing a value of 12 volts. This value exceeds the input so that bit is set to a 0 and the next bit to a 1 giving a value of 10 volts at the output of the DAC. This is still less than the input value and so the LSB is set. However, there is now a conflict because the next test shows that the input voltage is less than 11 volts. The binary value which is presented at the output is therefore 1010 (10 in decimal). Our simple 4-bit example does not have the resolution to cope with the

*ELECTRONIC MEASUREMENT TECHNIQUES*

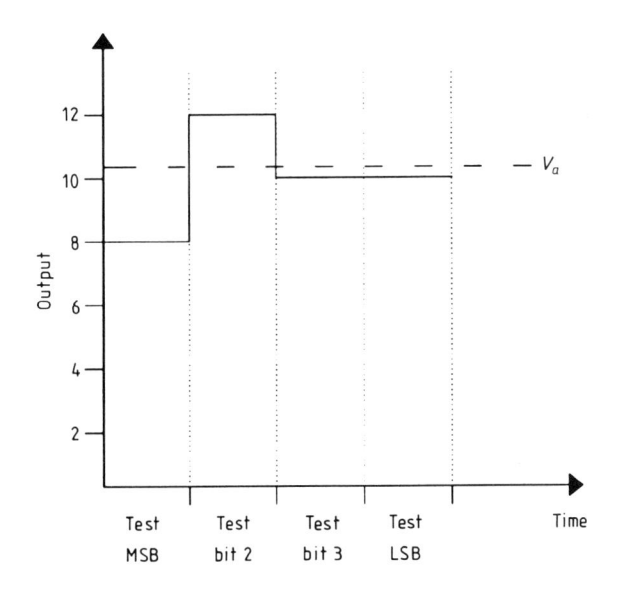

**Figure 4.14** A graph showing the successive outputs from the DAC for a 4-bit successive approximation ADC.

'extra' 0.3 volts as it can only distinguish voltage steps of greater than 1 volt. This difference is the quantization error (§ 4.5).

In general, an $n$-bit successive approximation ADC requires $n$ steps before the output value is obtained. Hence this type of ADC tends to be relatively fast, and a conversion time of around 10 μs is typical. Commercial devices usually have between 8 and 12-bits resolution. If a fast changing analogue waveform is to be digitized, it usual to include a S/H device to keep the input constant during the conversion time. High resolution successive approximation ADCs, of 16-bits or more, tend to be expensive. The dual slope integrating ADC tends to be more popular and cheaper if speed is not important.

*4.3.2.3 Dual slope ADCs*
This technique provides an ADC of very good accuracy without putting extreme requirements on individual components. One possible configuration is shown in figure 4.15. The main elements are an integrator, a zero crossing detector and a binary counter together with some logic gates and analogue switches. The system starts with $V_a$ connected to the integrator, which is reset together with the binary counter. The integrator is then enabled by pulling the reset line low, hence causing the integrator output to ramp negative. At this stage the output of the zero crossing detector will be high thereby allowing the clock signal to increment the binary counter. When the count is maximum, the overflow becomes high and -$V_{ref}$ is connected to the integrator. Effectively, the input voltage has been integrated over

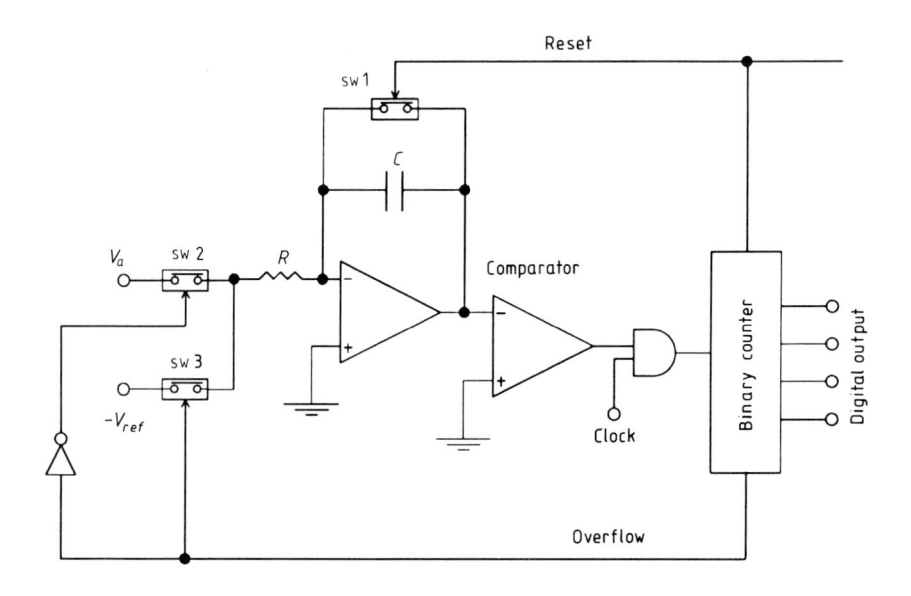

**Figure 4.15** A dual slope integrating ADC.

a known period of time, i.e. the maximum number of binary counts. With $-V_{ref}$ connected to the integrator, it ramps positive until zero is reached. The number on the binary counter is now proportional to $V_a$.

Figure 4.16 shows the integrator waveform during one conversion cycle. Denoting $t_a$ as the time between $t_o$ and $t_1$ and $t_{ref}$ as the time between $t_1$ and $t_2$ and assuming that the integrator has a time constant equal to the product $RC$, then in general the integrator output $V_o$ is given by

$$V_o = \frac{-tV_a}{CR}. \quad (4.16)$$

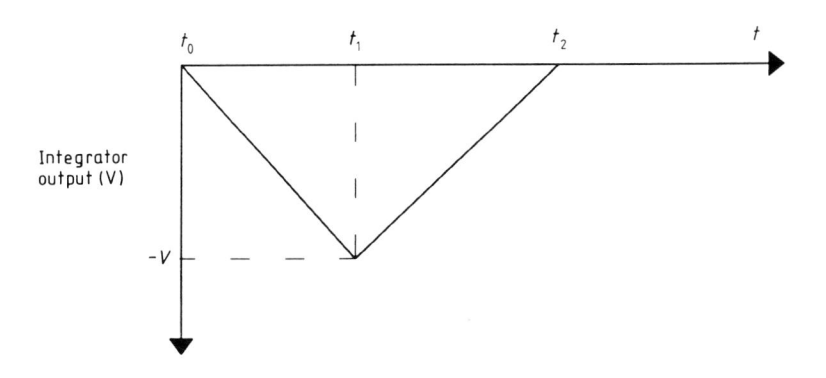

[Figure: Integrator output waveform — triangular shape from $t_0$ down to $-V$ at $t_1$, back up crossing zero at $t_2$.]

**Figure 4.16** The output waveform of the integrator during a conversion.

Now,

$$t_a = \frac{2^n \text{ counts}}{\text{clock rate}} \qquad (4.17)$$

where $n$ is the number of stages in the counter, i.e. the resolution of the ADC. Also

$$t_{ref} = \frac{\text{digital magnitude}}{\text{clock rate}}. \qquad (4.18)$$

The change in the integrator output during $t_a$ is $-V_1$ and the change during $t_{ref}$ is $V_1$. Hence

$$\frac{t_a V_a}{CR} = \frac{t_{ref} V_{ref}}{CR} \qquad (4.19)$$

substituting the expressions for $t_a$ and $t_{ref}$ we have

$$\text{digital magnitude} = \left(\frac{V_a}{V_{ref}}\right) 2^n \qquad (4.20)$$

so the output is not dependent on the time constant $CR$. Owing to the integration of the input waveform, any high frequency noise is filtered. Dual slope integrating ADCs are capable of giving very high resolution, but the main drawback is that the conversion speed is relatively slow. Typically the conversion rate is less than 30 per second. Some types of dual slope converter use an automatic zeroing technique. Here the input is short-circuited before the measurement cycle so that any output over the integration time must be zero-drift error, which can then be subtracted from the following reading, to give a continuous zero correction facility. In general terms, dual slope integrating ADCs offer the highest resolution (typically 10- to 16-bit) per unit cost and are widely used in applications such as digital multimeters, where the conversion speed is not so important.

### 4.3.2.4 *Parallel encoder ADCs*

One of the fastest types of ADC is the parallel encoder, or flash converter. This device is shown in figure 4.17. The analogue input is simultaneously applied to the input of each comparator. The comparators have equally spaced thresholds, generated by by the voltage divider chain of resistors. For a given input, the comparator outputs have a distinct pattern, with low outputs for those with thresholds above the input voltage, and high outputs for those with thresholds below the input voltage. The delay time is the sum of the delay of the comparator and the encoder. Speeds of up to 30 million samples per second are possible.

As the number of bits increases, they become prohibitively expensive and bulky as the priority encoder becomes more complex. The number of comparators

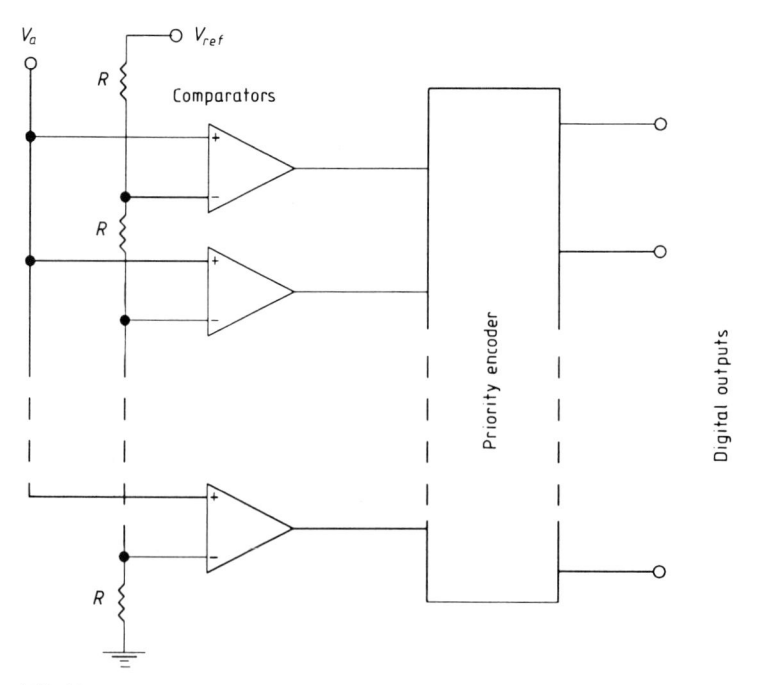

**Figure 4.17** Flash (parallel encoder) ADC.

approximately doubles for each additional bit. Commercial devices are usually available in the range 4-bit to 10-bit resolution.

This completes our brief review of the main techniques of data conversion. The important points to note are the usual engineering trade-offs between speed, resolution and cost. As a footnote it is worth remembering that a situation may arise where commercially available devices may not provide the optimum solution. For example, the new generation of intelligent sensors makes use of Application Specific Integrated Circuits (ASICS) whose standard cells may not include data conversion elements with the right combination of resolution and speed. In this instance the sensor designer could find himself involved in the development of new cells, like ADCs, to add to the existing cell library.

### 4.3.3 Gain control

The question of how to set the gain of the data acquisition system is far from the simple task it was in early days, and in intelligent sensor systems it can be one of the most difficult design aspects. Gain setting requires a delicate balance between utilization of data capacity (§1.2), quantization noise (§ 4.5) and the possibility of data loss at the tails of the signal density function (§ 2.1.8). If the gain is too low the information capacity of the data word is being wasted and the signal to quantization noise ratio is at a level that becomes unacceptable. If it is too high information will be lost through channel limiting. Furthermore the trade-off is highly case sensitive, so it is not easy to produce a common rule.

Consider the case of a sub-system comprising a gain-controlled amplifier followed by an 8-bit ADC. If the sampling process produces the number 254 from the converter, then we know within the accuracy of the system that this represents the value of the external variable. If, however, it produces the number 255 then we know that the external variable corresponds to 255 or any other number greater than 255. Now we have not lost all of the information in this case, because we know that the signal level is not less than 255, but we have lost some of it, and whether this loss is destructive depends upon the application. To take an everyday example, in a music channel any such distortion would be intolerable, while in a voice channel it might to some extent be tolerated. Conversely if the gain is so low that the signal only occupies the bottom 3-bits, the consequent signal to noise ratio due to quantization is only 18 dB (§ 4.5), which is generally unacceptable. Note that subsequent digital gain (multiplication by a constant) does nothing to enhance this ratio.

Signals from the real world are often unbounded, i.e. it is not possible to define an upper or lower limit that they cannot pass. We would not expect to observe a one megavolt noise peak from a one ohm resistor, any more than we would expect to see a twenty foot man, but the mathematical distributions which model these populations decay continuously at their tails, and while the numbers they produce for the cases mentioned are incredibly small they are still finite. This is a difficult area of mathematical philosophy, and certainly beyond the scope of this text; yet the decision on how much of the tails of the signal density function to snip off at the data acquisition stage is a very practical one, which we have to face in many applications of instrumentation.

In many areas of sensing the upper and lower bounds do exist (e.g. the contents of a tank). In many others they do not, which is generally the case for natural phenomena (rainfall, temperature, atmospheric pressure etc.). There is a special branch of statistics, the statistics of extremes (Gumbel 1958 ), which deals with the analysis of these events at the tails of distributions, to which we would have to resort for an accurate analysis of the probability of data loss. The general problem is to strike a balance between data loss at the extremes and quantization errors at the median. This means that for many applications the numbers around maximum and minimum from our ADC are rarely used, which is considerable deviation from the equiprobable case which would give us maximum source entropy (§ 1.2). One of the techniques available in intelligent sensor systems is computed automatic gain control, but this has to used with great care, especially in closed loop control systems. The trade-off can become quite complicated. The requirement to minimize data loss calls for the gain to be set low, while the dual requirements for efficient use of the channel of information and minimization of relative quantization errors call for it to be set high.

## 4.4 THE EFFECT OF NOISE ON ANALOGUE SYSTEMS

Noise is an unwanted intrusion in any system and is generally considered to be any signal that obscures the desired one. The noise may be of physical origin arising from the components used within the system, or it could be an external signal (interference). Noise is usually characterized in a number of ways including its frequency spectrum, amplitude distribution (§ 2.1.8) and the physical mechanism responsible for its generation. In the following sections we shall describe the most important noise classes.

### 4.4.1 Thermal noise

This type of noise arises from the random motion of charge carriers within resistive materials. Thermal noise is random in nature and exhibits a flat spectrum, for which the term *white noise* is often used. An expression for the (RMS) noise voltage generated in a resistance $R$ at temperature $T$ is given by

$$V_n = ( 4kTBR )^{1/2} \tag{4.21}$$

where $k$ is Boltzmann's constant, $T$ is the absolute temperature and $B$ is the bandwidth. For a 10 kΩ resistor at room temperature, measured over a bandwidth of 10 kHz, the noise voltage would be about 1.3 µV (RMS).

### 4.4.2 Shot noise

This is the characteristic noise occurring in semiconductor components like diodes and transistors. An electric current is produced by the motion of discrete charge carriers, namely electrons and holes. Consequently direct current, rather than being smooth and continuous, is the sum of numerous small pulses caused by the charge carriers. An expression for the (RMS) current noise in a semiconductor device is given by

$$I_n = ( 2qI_{dc}B )^{1/2} \qquad (4.22)$$

where $q$ is the electronic charge, $B$ is the measurement bandwidth and $I_{dc}$ is the steady current. The formula for the shot noise assumes that all charge carriers have the same charge and that the motion of a single charge carrier is statistically independent of any other charge carrier. Like thermal noise it is white in nature.

### 4.4.3 $1/f$ noise

$1/f$ noise (flicker noise) is an almost ubiquitous type of noise originating from a number of mechanisms in electronic devices. It is generally considered to arise from imperfections in the manufacturing process of the components. The term *excess* noise is often used to describe this type of noise because it appears in addition to the white noise. The spectral density function for noise of this kind is characterised by an approximate $1/f$ dependence in the range of interest. At frequencies higher than a few hundred kilohertz, the $1/f$ noise amplitude is small and white noise tends to dominate. $1/f$ noise is particularly prevalent in devices which rely upon surface conduction, as the degree of imperfection in the surface of the material tends to be greater than that in the bulk. A notable example here is the case of MOS transistors where conduction is largely over the surface. However, in JFET and bipolar transistors the conduction is mainly beneath the surface. The consequence here is that MOSFETs exhibit a higher $1/f$ noise content and tend to be avoided for low frequency applications. $1/f$ noise can be observed in some types of resistor. If a steady current is passed through a carbon resistor, then a voltage fluctuation can be measured across its terminals. Many observers suggest that this arises from a fluctuation in the number of charge carriers within the device. The effect is barely noticeable in high quality metal film and wire-wound resistors. $1/f$ noise has also been observed in nature in such diverse examples as the speed of the ocean currents, the flow of sand in an hour-glass.

### 4.4.4 Interference, screening, shielding, and grounding

Electrical interference is a very common problem encountered in measurement systems. The problem is associated with coupling by conduction, capacitive links, mutual inductance or by radiation via stray capacitances and mutual inductance. An important potential source of electrical interference arises if the measurement system is connected to ground at more than one point, which results in *ground loops*.

Electrostatic coupling exists between any two conductors as a result of the (stray) capacitance between them. A varying electric field will cause an induced interfering voltage on a signal wire. One simple remedy is to ensure that the measurement circuit is not positioned close to power lines, motors, fluorescent lights etc. This problem is particularly prevalent with high impedance circuits, so as a general rule, low impedance transducers are preferred where other trade-offs permit. For relatively low frequency circuits, screened cable can be used to minimize the problem of capacitive coupling. The signal conductors are surrounded by a metallic shield which is grounded at one end. Any induced charges are therefore coupled to ground and do not appear on the signal conductors. For high frequency applications coaxial cable is used to minimize this problem.

A conductor which is carrying an alternating current is surrounded by an alternating magnetic field. Any conducting material placed within this field will therefore experience an induced voltage. One way to overcome this effect is to completely surround the circuit with a magnetic shield so that a high permeability path exists to divert any interfering magnetic flux. An alternative approach is to twist the two signal conductors together so that the interfering flux linkages are effectively cancelled out. Commercially available cables with twisted wires and wrapped foil shields are available to meet the needs of both electrostatic and inductive noise reduction. It is generally recommended that cables are properly secured in order to eliminate the so-called *triboelectric* effect which is associated with rubbing friction within the cable itself and can induce noise voltages.

The ground loop problem mentioned above is a major problem in lax design, which arises because ground points within a circuit or system are seldom at equal potentials. The conducting path which is serving as the ground generally has a some resistance and carries both intentional and spurious currents. Single point grounding, preferably at the signal source end, is recommended. However, other factors also affect grounding policy, safety being a very important one, and in many systems, particularly large ones, optical or inductive isolation is imperative.

### 4.4.5 Recovering the signal

Many methods exist for electronic recovery of the signal from the noise and anyone who is not familiar with these techniques is advised to consult existing texts which cover this area in more detail than we can provide here (Wilmshurst 1985, Horrowitz and Hill 1990). However, it is worth reminding ourselves of three important techniques which are appropriate to to the concepts of intelligent sensor systems.

#### 4.4.5.1 *Filtering*

This may take the form of anything from a traditional analogue filter, using a simple *RC* network as a low-pass filter to remove high frequency, random noise, to a a complex digital filter capable of tailoring its characteristics to match those of the system (§6.4.3.). Analogue filters may be either **passive** or **active**. Passive filters require no external energy source, the *RC* network being a good example. Active filters utilize the advantages of op-amps and are thus able to provide electrical filters which are very versatile and can cover a wide frequency range.

In terms of noise filtering, there are essentially four classes of frequency selective filter which pass the desired signals and reject the spurious ones. **Low-pass** filters allow a band of frequencies below a defined cut-off point to pass through the filter, ideally without experiencing any attenuation. All signals of frequency greater than the cut-off frequency are severely attenuated and effectively rejected. All practical filters differ from this idealized response which is sometimes referred to as a *brick wall* or *block function* response. Rather than exhibiting a perfect cut-off characteristic, real filters have a transition phase in their frequency response from the pass band to the stop band. Various filter realizations exist which are able to approach the ideal response, but inevitably, there are trade-offs with other factors such as response time, circuit complexity, pass band characteristics etc. **High-pass** filters have the ability to pass high frequencies and reject low frequencies. As an example, if a situation arose where the mains frequency was found to be interfering with the signal from a transducer, then a high-pass filter with a cut-off frequency above 50-60 Hz will alleviate the problem provided that the frequency range of interest is well beyond this region.

**Band-pass** filters are used when there is a need for a particular range of frequencies to be transmitted and all other frequencies to be rejected. By cascading a low-pass and high-pass filter, a band-pass response can be achieved. Sharply tuned band-pass filters are used in the design of audio mixers and spectrum analysers. Returning to the problem of eliminating the mains frequency from a transducer signal, the disadvantage of using a high-pass filter is that all frequencies below the cut-off will be rejected. It would be preferable to remove only the 50-60 Hz frequency band. This can be achieved with a **notch** or band-rejection filter, which can be employed with minimum effect on the desired signals.

Filtering may also be performed digitally using a computer. Digital filters generally offer greater flexibility, in terms of the variety of applications, than their analogue counterparts. The basic concept of the digital filter was introduced in Chapter 2 and will not be elaborated upon here as they will be discussed further in § 6.3.1 and a specific application in an intelligent sensor is covered in § 6.4.3.

#### 4.4.5.2 *Signal averaging*

A relatively simple way of eliminating random noise from a signal using a digital method is by multiple time averaging. If $m$ samples of a noise-contaminated signal are stored and averaged, using, say, a computer, then an improvement in the signal-to-noise ratio can be observed. Suppose a signal $x$ can be represented as the sum of a digitized signal $S$, and a digitized noise component $N$ then,

$$x(t_k + iT) = S(t_k + iT) + N(t_k + iT) \tag{4.23}$$

where $t_k$ is the time at which the $k^{th}$ sample is taken, $i$ is the sample being observed, $k$ is the number of sweeps of the waveform and $T$ is the sampling time. This can be implemented as a recursion (equation (3.2)). If the signal is periodic and is sampled at the same point for each sweep then

$$S(t_k + iT) = (iT). \tag{4.24}$$

Assuming that the noise component is random and has zero mean value, then the RMS value of the noise is $\sigma$. If we consider the summation of $m$ values then,

$$\sum_{k=1}^{m} x(t_k + iT) = \sum_{k=1}^{m} S(iT) + \sum_{k=1}^{m} N(t_k + iT). \tag{4.25}$$

The mean square value for the sum of $m$ noise samples = $m\sigma^2$, and hence the RMS value of the noise is $\sigma\sqrt{m}$. The signal-to-noise ratio is therefore,

$$\left(\frac{S}{N}\right)_m = \frac{mS(iT)}{\sigma\sqrt{m}} = \sqrt{m}\left(\frac{S}{N}\right) \tag{4.26}$$

so under these conditions it can be seen that there is an improvement in the signal-to-noise ratio by a factor of $\sqrt{m}$, indicating that by averaging over a greater number of traces, the noise component can be reduced.

Noise reduction is an important activity in sensor applications, and where it is relevant the signal averaging process can be accomplished by a simple recursion, as we saw in our discussion of noise as a defect (§3.2).

### 4.4.5.3 *Correlation*

The correlation techniques discussed in Chapter 2 can be used for recovering a signal from one contaminated with noise. An example here is the recovery of an ECG signal generated by the heart (Wilmshurst 1985). This signal is periodic but the exact shape of the beat can appear distorted due to noise within the system. Autocorrelation techniques have been successfully adopted for recovering the signal. The autocorrelation function of a stationary signal $x(t)$ is defined in time average form (§ 2.1.8) as

$$R_{xx}(\tau) = \lim_{T \to \infty} \frac{1}{2T} \int_{-T}^{T} x(t)\, x(t + \tau)\, dt \qquad (4.27)$$

where $\tau$ is a variable time delay. The value of the autocorrelation function is usually expressed graphically in the form of an auto-correllogram, an estimate of the autocorrelation function, which is essentially a plot of $R_{xx}(\tau)$ against $\tau$. Auto-correllograms can reveal hidden periodicity in what appears to be a random signal. The auto-correllogram of wide band random noise appears as a narrow pulse centred around $\tau = 0$ and decaying rapidly with increasing $\tau$. The autocorrelation function of a periodic function such as a sine wave is itself periodic, hence a noisy sine wave would appear largely as a sine wave in the auto-correllogram but with distortion around $\tau = 0$.

## 4.5 QUANTIZATION AND OTHER DIGITAL NOISE SOURCES

### 4.5.0 Introduction

Sensor applications require the digital processor to perform operations on signals which are derived from the outside world. In general, these signals will be continuous variables, whereas the operation of the processor itself is discrete and sequential. The continuous variable may be presented to the processor by means of analogue-to-digital conversion (§ 4.3.2). Such a conversion, however, introduces quantization errors. The degree of error introduced by quantization is a direct consequence of the nature of the process of analogue-to-digital conversion. The factors which now become important when the microprocessor performs manipulations with the quantized signal are those of linearity, noise, and stability. These aspects are considered in the following discussion.

Mathematically the continuous to discrete conversion is a many- one mapping, which is essentially non-linear, i.e. each value on the discrete side maps to an infinite number of values on the continuous side.

### 4.5.1 Fundamental non-linearity of quantization

A linear system is one which is additive and homogeneous (figure 2.2). No real system is totally linear but within a certain range of signal magnitudes and frequency limits it can be very nearly so. The linearity criterion is strictly applicable only to continuous variables, and can be applied to discrete systems only if the number of values of variables is restricted to a finite sub-set.

When continuous voltages are converted into digital numbers there are residual errors, as we shall see in the analysis below. If two voltages are added together the results can be different from the addition of the two corresponding numbers should the errors sum to the value of the least significant bit. In this respect the analogue to digital converter is non-linear. Such systems are only capable of quasi-linear operation in that they are additive and homogeneous within certain quantization approximations and because of the lack of a general non-linear theory we tend to avoid this aspect and pretend we are dealing with a noise problem. The difference between the analogue input signal and quantized output signal constitutes this quantization noise.

### 4.5.2 Treatment as a noise problem

As we do not have a general non-linear theory we find it convenient to treat quantization as though it were a different problem entirely, i.e. one of noise. How this is done can be seen in figure 4.18. Figure 4.18 (a) represents an arbitrary waveform, whose quantized version is shown in figure 4.18 (b). Now, though this transformation has been wrought by a non-linear process, exactly the same result could have been achieved by adding a signal, namely that shown in figure 4.18 (c). By examining the properties of this imaginary signal we are able to obtain useful results for the assessment of the effects of quantization.

Looking at the quantization noise in figure 4.18 (c) we observe that it is largely made up of triangles of equal amplitude. This observation does not apply so well at the turning points, but the errors are relatively small and tend to be self cancelling. Now we have already examined triangular waves in an earlier chapter and have established some statistical properties. If the range of the input voltage is $V$ and the number of bits of conversion is $n$, then we can identify a number of quantization levels $q$, where $q = 2^n$, and the quantization gap is $\Delta = V/(q-1)$, which is also the amplitude of the triangular waves of quantization noise as observed in figure 4.18 (c). We have already established (§ 2.1.8) that the mean value of the quantization noise is then $\Delta/2$ and the mean square is $\Delta^2/3$. Furthermore removing the zero frequency component (i.e. the mean) we found that the mean square amplitude is given by $\overline{N^2} = \Delta^2/12$. This is termed the noise power, and in some older texts is expressed in watts, as with $\Delta$ in volts it would be the power dissipated by the quantization noise in a one ohm resistor.

*ELECTRONIC MEASUREMENT TECHNIQUES* 117

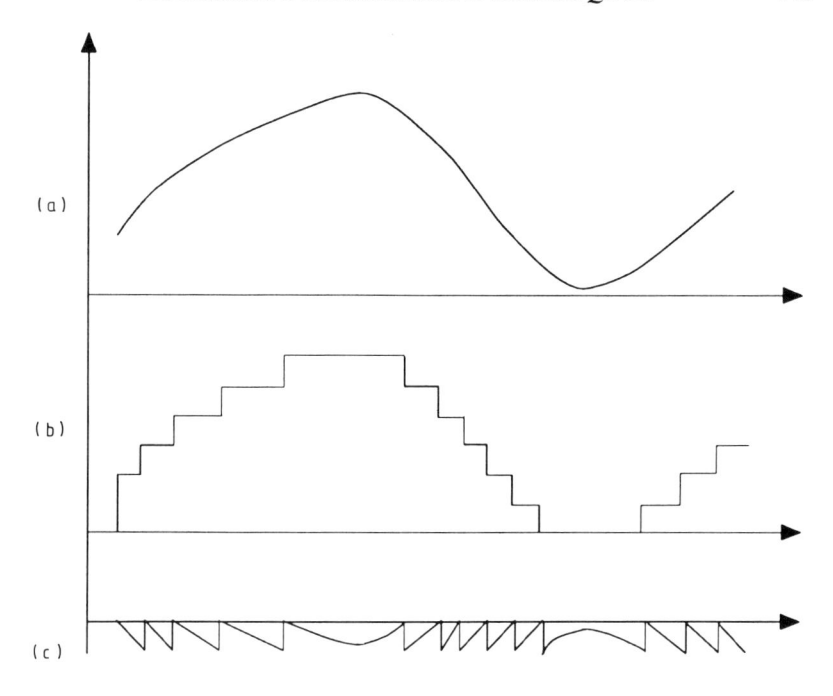

**Figure 4.18** (a) an arbitrary section of input signal; (b) its quantized version and (c) the difference, which represents the hypothetical quantization noise.

### 4.5.3 Importance of word length

Let us make a simplifying assumption that the actual signal is uniformly distributed (e.g. our familiar triangular wave) and is of amplitude $V$. By the same argument as above we can assert that the signal power is given by

$$\overline{S^2} = \frac{V^2}{12} = (q-1)^2 \frac{\Delta^2}{12}. \tag{4.28}$$

Thus the signal to noise ratio is given by

$$\left(\frac{\overline{S^2}}{\overline{N^2}}\right) = (q-1)^2 = (2^n - 1)^2 \approx 2^{2n}. \tag{4.29}$$

Such ratios are normally expressed in dB, so we have

$$\left(\frac{\overline{S^2}}{\overline{N^2}}\right)_{dB} = 10\ log_{10}\ 2^{2n} = 20n\ log_{10}\ 2 \approx 6n \tag{4.30}$$

Thus we have the very useful rule of thumb that the signal to noise ratio is six times the word length in bits, but only if we use the whole range of the word. Otherwise it is six times the number of bits spanned by the signal. Hence the importance given to the problem of gain setting (§ 4.3.3).

### 4.5.4 Possibility of amplification of noise in digital systems

Subsequent manipulations of the quantized signal may cause an increase in the effective quantization noise in the signal. In particular, arithmetic processes which mimic differentiation and integration, while discrete processes with resonant characteristics can also cause problems. Such mathematical operations must thus be used with care. This can be particularly important if the processed signals are returned to the real continuous world, and possibly devastatingly so if this involves a closed loop.

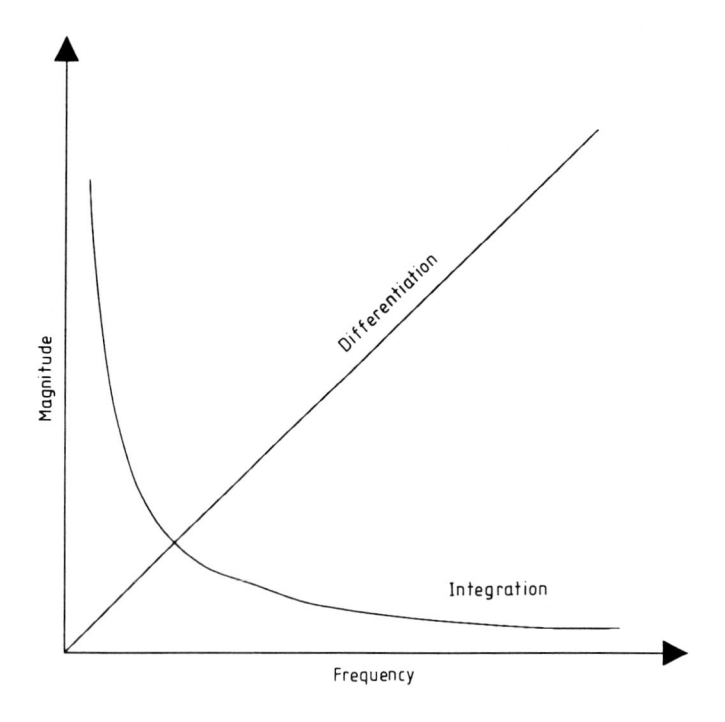

**Figure 4.19** The magnitudes of the operators differentiation and integration plotted as a function of frequency.

The amplification of noise by the processes of differentiation and integration may be considered by observing the equivalence of these operations from the time domain to the Laplace domain, and to the Fourier domain. Thus for differentiation,

$$\frac{d}{dt} \rightarrow s \rightarrow j\omega \qquad (4.31)$$

and for integration

$$\int () \, dt \rightarrow \frac{1}{s} \rightarrow \frac{1}{j\omega} \, . \qquad (4.32)$$

Differentiation thus tends to amplify high frequency noise while integration tends to amplify low frequency noise. The effect on signal frequency by these two mathematical operations can be seen by plotting the magnitude of the corresponding operator as shown in figure 4.19.

The unwanted low frequency components tend to come into the system from the outside world or from the primary sensor, largely in the form of drift or $1/f$ noise. The unwanted high frequency components, on the other hand come not only from the outside but also from the conversion and arithmetic processes; so differentiation, or its digital analogues, can cause amplification of signals that are entirely spurious as far as the real world is concerned. To illustrate this let us apply an approximate form of differentiation, the first difference (§ 6.3.2), to the quantized version of our signal from figure 4.18 (b) This is, of course, a very crudely quantized signal, but it is the sort of thing that can easily occur at low levels of the input signal, especially if we have tried to apply the rather dangerous process of automatic gain control within the digital process (§ 4.3.3). As the quantized signal is in the form of of an irregular staircase function it is clear that the first difference will be in the form of a series of blocks, whose width depends on the interval over which the difference was taken, and whose heights are determined by the rises of the staircase. Thus the results of the operation, as we can see in figure 4.20, bear little relationship

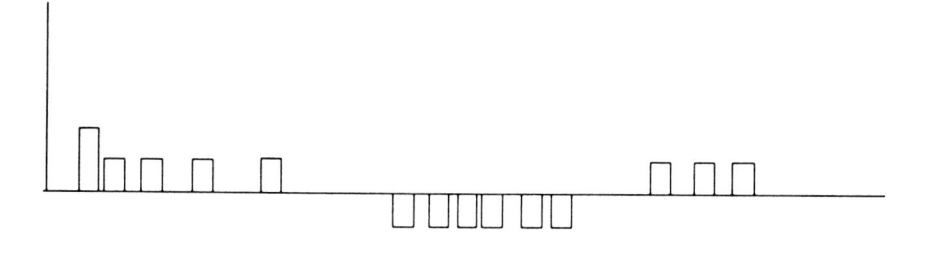

**Figure 4.20** Approximate differentiation as applied to the quantized waveform of figure 4.18 (b).

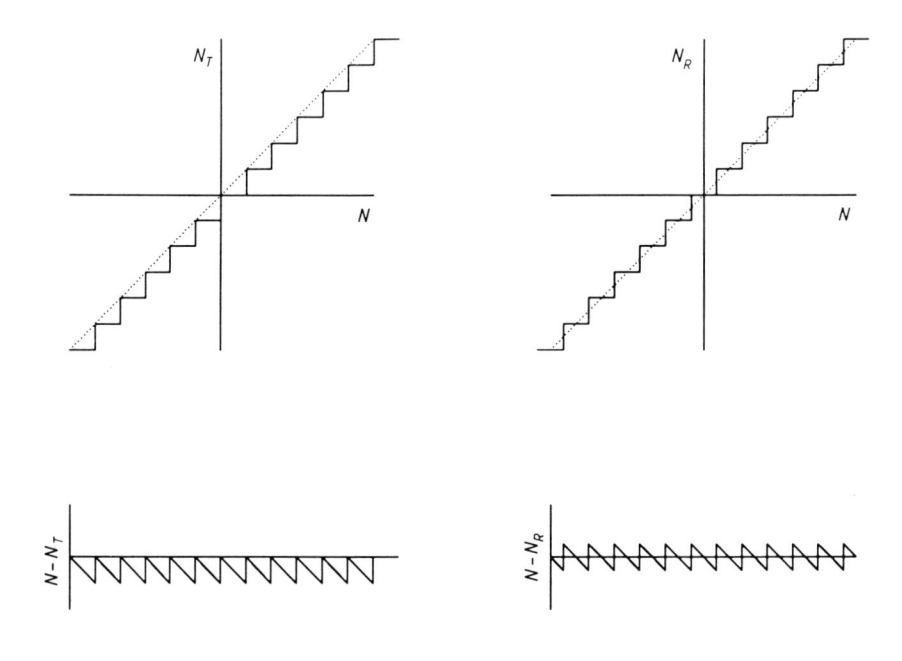

**Figure 4.21** Illustration of the processes of truncation and rounding. Below are the effective equivalent noise waveforms.

with the slope of the original signal in figure 4.18 and the only information that is preserved is effectively the sign of the slope. The dangers of using information processed in such a way are self evident, especially if we return them to the outside world through a DAC.

### 4.5.5 Truncation and rounding

Other effective digital noise sources that occur in the processing of numbers, especially in the fixed point processing of signals in processors of restricted word length. However, even if we have the luxury of floating point arithmetic internally, our numbers still have to be squeezed through an output DAC or a digital bus. There are two basic methods of shortening a number, truncation and rounding. In truncation bottom bits of the number are simply removed, while in rounding there is the additional step of adding a new LSB if the change so produced is greater than half the new LSB. These operations are best illustrated graphically, as in figure 4.21.

Truncation is the process normally applied in AD conversion as we have seen. The noise it produces is always negative and of the peak amplitude of one LSB. Rounding on the other hand produces noise with zero mean and a peak amplitude of one half LSB. Both have our familiar triangle waveform of § 4.5.3, and while the average error is smaller with rounding, the noise power, disregarding the mean, is the same in both cases.

# 5
# Enabling Technologies

## 5.0 INTRODUCTION

Selecting suitable methods of implementing intelligent sensors can prove to be a substantial task when one bears in mind factors like cost, robustness, size, packaging and performance. Advancements in many areas of microelectronics over the past decade or so have opened up many new fields of sensor implementation, the four major areas being silicon, thick-film, thin-film and optical fibre technologies. Each of these has its own advantages and disadvantages and often a hybrid approach utilizing a combination of these fabrication technologies may provide an effective solution.

On the evidence provided by the number of scientific papers published and patents, it is clear that the research undertaken in the areas of fibre optics and silicon technologies far outweighs that addressed to the film technologies. The printed circuit board (PCB) fabrication process has been successfully employed in the sensor industry for many years. For the purpose of this text and in view of the accepted status of PCB technology we shall classify this as a traditional technique of sensor methodology.

## 5.1 TRADITIONAL TECHNIQUES

The printed circuit board is a popular technology for sensor realization. The primary use of PCBs for sensor applications is in the provision of the support electronics. This includes stabilized excitation voltages, signal conditioning circuits and compensation hardware. Many different processes have been used to produce PCBs but the most common method is the *etched foil* process. A PCB essentially comprises a suitable substrate and conducting tracks which serve as both interconnecting lines and also lands for the subsequent addition of other electronic components. The choice of substrate usually depends on the application of the circuit and factors such as flexibility, expansion coefficient, reliability, and environmental conditions need to be considered. A common form of substrate is a glass fibre-reinforced plastic laminate.

The etched foil process is usually performed on copper-foil clad laminate substrates and is a subtractive process. The board is coated with a resist material in the configuration of the desired conductor pattern. The exposed copper is then removed by a suitable etchant like ferric chloride. The board is then washed in running water to remove any remaining etchant and then dried. The resist is then dissolved in a solvent or removed by vapour degreasing. The process is illustrated in figure 5.1.

Essentially there are three main methods for applying the resist to the substrate:

*Screening* involves deposition of the resist material by transfer through the open areas of a mask (or screen). See § 5.4 on thick-films for a description of screen printing.

*Photo-methods* rely on the properties of various wavelengths of light to harden a selective area of photosensitive resist solution. The board is exposed to the light through a photographically produced negative of the desired conductor pattern.

*Offset printing methods* are based on conventional lithographic printing techniques whereby the lithographic plate is designed only to accept ink on the printing area. The inked pattern is transferred to a cylindrical rubber blanket and then rolled onto the foil clad substrate. The resulting ink image is a good representation of the

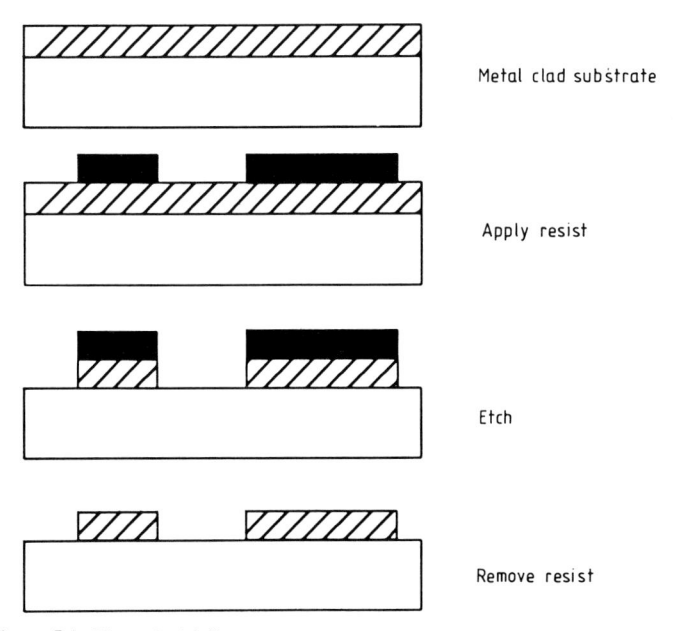

**Figure 5.1** The etched foil process.

original but the print thickness alone is not suitable. It is reinforced by dusting on and subsequently fusing a natural resin to act as the resist.

Sensor applications tend to require circuit boards with a high density of components because of the limited space within the housing. Hence double-sided PCBs tend to be used frequently, and a more elaborate through-hole process is needed. At each point on the top surface, where a connection needs to be made to the underside, a hole is drilled. If an etched foil process is to be employed the substrate takes the form of a double-clad board which is operated on in a similar manner to its single-sided counterpart. Connection through the board is accomplished using several different approaches. One way is simply to use a metal pin which can be soldered on each side of the board. This is a low-cost solution for small production runs but is rarely used when large quantities of PCBs are required. The etched foil through-hole method requires a smooth surface on the through-hole walls which are sensitized by a thin layer of electroless plated copper. A negative resist pattern is then deposited on both sides of the board thereby covering all the copper except for the areas to be used as conductor tracks or pads. A layer of copper is then deposited onto the hole walls and tracks by an electroplating technique, after which a thin coating of solder is applied. The solder acts as a metal resist for the further etching of the negative resist and unwanted copper.

Another through-hole process is the plated through-hole technique shown in figure 5.2. The first step involves drilling the required holes in an insulating substrate and then applying a thin adhesive layer to all the surfaces. This layer is not fully cured at this stage. A thin sensitized layer, usually copper, is then chemically deposited onto the adhesive layer. A negative photoresist is next deposited in proper registration onto both sides of the board. The required thickness of copper is then electro-deposited onto all exposed areas thus giving the desired conductor pattern together with the through-hole connections.

The resist coating, sensitized layer and exposed adhesive are removed by etching and the board is then subjected to a high temperature and pressure cycle to cure the adhesive and remove any remaining residues.

The subsequent processing steps depend on the application of the circuit. In most cases other components like resistors, capacitors, transistors, diodes etc., will need to be added and numerous techniques exist for soldering these components including vapour phase reflow, dip and roller coating methods. Of course, for relatively small circuits components may be soldered manually.

Another optional processing step is the addition of a passivation layer onto the circuit to protect it from the environment. Sometimes a solder resist layer is required and this coating, usually a thermosetting plastic, is applied by a screen printing technique. A variety of lacquers and varnishes are available which can be easily administered to the circuit board to provide protection from the environment. Other types of conformal epoxy coatings are also available and these are deposited by spraying, dipping or brushing, usually at room temperature.

Substrate with holes

Surface sensitized and
adhesive applied

Negative resist
deposited

Electrodeposition of
copper

Resist, exposed adhesive
and sensitized layer
removed

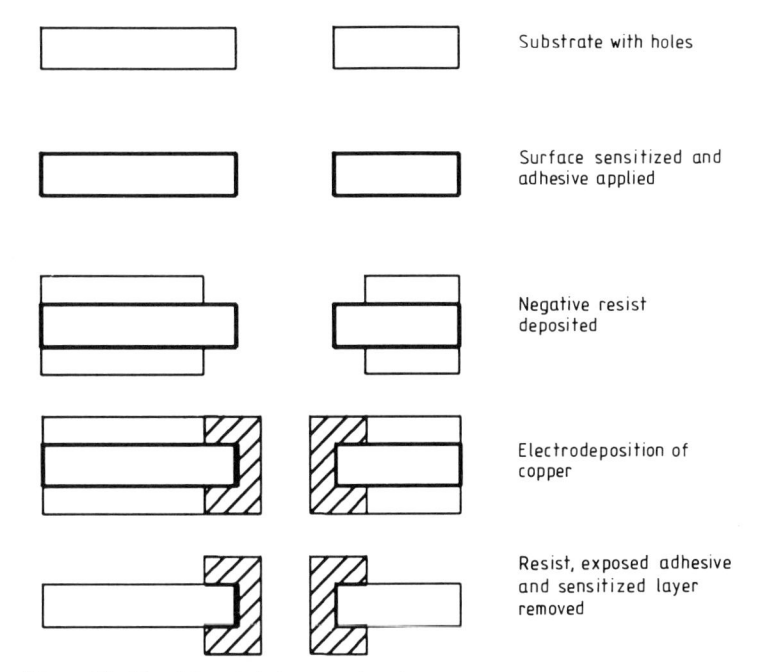

**Figure 5.2** Plated through-hole processing steps.

## 5.2 SILICON PLANAR TECHNOLOGY

### 5.2.1 Introduction

Silicon planar technology is certainly the most common processing method for silicon (Si) integrated circuits. The early semiconductor work was performed with germanium (Ge), but this was superseded by silicon because of its higher reliability, improved temperature capability and because silicon dioxide ($S_iO_2$) acts as a very good barrier and is easy to implement. The whole field of IC technology has arisen from the invention of the transistor in 1947. Until the 1960s the number of transistors which could be fabricated on one chip was limited to one, but throughout the past three decades we have seen that figure rise to around one million transistors per chip using today's Very Large Scale Integration (VLSI) techniques. Future predictions estimate that by the end of this century, it will be possible to produce a single silicon chip with a population of over 10 million transistors.

The two fundamental silicon device technologies are bipolar and Metal Oxide Semiconductors (MOS), both of which are fabricated using a planar technology indicating that the manufacturing processes all take place in a single plane. The importance of silicon device technology to sensor development became evident in the early 1970s with the introduction of the ISFET. In this section a general review of the basic silicon planar technology processing steps will be presented together with a coverage of bipolar and MOS devices. The area of micromechanical sensors has attracted great interest recently and a brief resumé of these devices will also be covered.

### 5.2.2 The silicon planar process

Silicon planar processing is by far the most highly developed fabrication method compared to that associated with other semiconductor materials. Silicon does not posses ideal properties in many respects. Gallium arsenide allows improved frequency response due to higher electron drift mobilities but the processing capabilities of this material are more complex and less perfected than that of silicon.

#### 5.2.2.1 *Crystal Growth*

The fundamental starting point of the process is the silicon substrate. $SiO_2$ (sand) provides the basic ingredient. The Czochralski (CZ) process is usually adopted to manufacture the substrate. This technique involves inserting a small seed crystal of silicon into a crucible of molten silicon containing the appropriate dopant such as boron or phosphorous. As the crystal is slowly pulled out of the melt, rotation speed and melt temperature affect the diameter of the resulting crystal. A typical crystal might be 10 cm diameter and 50 cm long. This is then sliced into wafers around 250 µm thick and subsequently lapped and polished to provide a suitable finish for subsequent processing.

#### 5.2.2.2 *Epitaxial Growth*

The word *epitaxy* is derived from the Greek words *epis* and *taxis* meaning layered and ordered. The epitaxial process allows the growth of single crystal silicon upon a crystalline substrate and is carried out in a furnace or reactor. The temperature in the reactor can be as high as 1000 °C. The process itself relies upon the oxidation of gases such as silicon tetrachloride ($SiCl_4$). Dopants may also be added during deposition. Typical thicknesses of epitaxially grown layers are between 2 µm and 20 µm. Epitaxial layers are used in bipolar and MOS technologies, and one example is as a buried layer in an *n-p-n* bipolar transistor. The method of gaseous deposition described is sometimes called vapour-phase epitaxy (VPE) but other techniques like liquid-phase epitaxy (LPE) and molecular-beam epitaxy (MBE) are also used and are covered elsewhere (Pucknell and Eshraghian 1987).

#### 5.2.2.3 *Oxidation*

One of the main attractions of silicon technology is the ability to grow an oxide ($SiO_2$) onto the silicon surface to act as a passivation layer. Thermal oxidation is carried out in an oxygen or water-vapour atmosphere at temperatures between 700 °C and 1200 °C. The thickness of the films is between 0.02 µm and 2 µm and these may also serve as either masks in the diffusion or ion-implantation steps to prevent dopant penetration, or as gate oxide films. The chemical equations for the oxidation processes are:

$$Si + 2H_2O \rightarrow SiO_2 + 2H_2 \quad \text{(wet oxidation)} \tag{5.1}$$

$$Si + O_2 \rightarrow SiO_2 \quad \text{(dry oxidation)} \tag{5.2}$$

### 5.2.2.4 *Photolithography*
The photolithographic process used in silicon technology translates a geometric pattern from a glass plate (or mask) onto the silicon substrate. The artwork is usually generated on a computer, using suitable layout software, and the working template is produced on a mask generator. One mask is required for each layer of the device. Electron-beam direct-write machines do not use masks but write directly onto the silicon wafer. Electron-beam also has the advantage of having a smaller wavelength than light, thereby allowing an improved line definition.

The wafer is uniformly coated with a photoresist, and the mask, usually a negative image, is aligned over this layer. Upon exposure to a UV source the exposed areas of the photoresist are polymerized and the unexposed areas are then removed during the development process. This process uses a negative photoresist, but it is also possible to use a positive photoresist polymer in which the exposed area is removed. Etching techniques are then used to generate the required pattern.

### 5.2.2.5 *Etching*
Various etching techniques are used to remove selected materials from the silicon wafer. The processes are characterized by their **selectivity,** i.e. the effect of the etch upon different substances, and the degree of **anisotropy.** Anisotropic etching occurs in only one direction (usually vertical) in contrast to isotropic etching in which material removed at the same rate in all directions. It is often undesirable to use an isotropic etch as this leads to under-etching of the mask pattern.

Wet etching involves exposing the silicon to chemical solvents which dissolve the unwanted areas. Essentially this is an isotropic process and may result in lateral deformation of the required pattern. Dry etching methods are very popular because of their anistropy. With this process the etching occurs under reduced pressure in the presence of a gas or plasma. Dry etching processes allow a high resolution pattern definition and smaller feature size than the wet-etch techniques.

### 5.2.2.6 *Diffusion*
Diffusion is a process in which impurity atoms move within a solid at elevated temperatures provided that a concentration gradient exists. On average the atoms

move in the direction of decreasing concentration. This technique is the most common method of producing *p-n* junctions. Dopants used include phosphorus, arsenic and antimony for *n*-type doping and boron for *p*-type doping. These materials move through the silicon by effectively hopping from one lattice site to another. The following equation describes the process,

$$\frac{dN(x,t)}{dt} = D\frac{d^2 N(x,t)}{dx^2} \tag{5.3}$$

where $D$ is the diffusion coefficient, $N$ the volume concentration, $x$ the distance and $t$ is time. Diffusion usually takes place through windows in the silicon dioxide layer overlying the silicon substrate. The diffusion process is actually non-linear and the diffusion coefficient is a function of the doping level. Most laboratories compensate for this by making use of calibration sheets based on experimental data.

#### 5.2.2.7 *Ion Implantation*
The second method for introducing impurities into the silicon is ion implantation. This process is performed in a vacuum at a lower processing temperature than that used with the diffusion technique. A beam of dopant ions is accelerated to an energy between 50 and 500 keV and directed toward the silicon substrate. The impinging ions have enough energy to enter the silicon surface to a depth somewhere between 10 nm and 1 μm. Silicon dioxide acts as a good barrier material and is frequently used as a mask. The two major advantages of ion implantation are the decreased spreading of the doped region and improved control over the doping profile.

#### 5.2.2.7 *Metallization*
The interconnection tracks and external connection contacts are formed by low-pressure deposition of metallic films. Aluminium and gold are commonly used as these posses a high conductivity and have the ability to form low-resistance ohmic contacts. Aluminium is probably the easiest material to work with but it suffers from a problem of migration, whereby the aluminium atoms migrate in the direction of the electron flow. This condition is worsened in a high humidity atmosphere. The problem can be minimized by the addition of copper. The required interconnection pattern is formed once again by the use of a mask and any unwanted film material is etched and removed.

In this sub-section we have attempted to give a very brief introduction to some of the basic processing steps used in the production of silicon integrated circuits. The interested reader is advised to consult the references given for a more in-depth approach to this subject, in particular, the text by Middlehoek and Audet (1989) which covers the field of silicon sensors.

### 5.2.3 Silicon micromachining

Having previously described silicon integrated circuit technology as a planar process we will now contradict ourselves and describe some techniques which allow the construction of three dimensional geometries for state-of-the art silicon transducers. Isotropic and anisotropic etching techniques are used to define the structure. Both micromachined sensors and actuators are realizable using these processes. One of the most common elements in micromachined sensors is the resonating beam, usually fabricated in the form of a cantilever. This simple structure can provide the basis for numerous types of mechanical measurement including acceleration, pressure, force and viscosity.

#### 5.2.3.1 *Etchants*
Common isotropic etchants mostly comprise a mixture of several acids like hydrofluoric, nitric and acetic acids. Generally, isotropic etchants are not preferred, as factors such as selectivity and precision are difficult to control. Small window areas take longer to etch than wider windows as their access to fresh etchant is restricted.

Anisotropic etchants exhibit a preferential etch rate toward certain crystals orientations. Potassium hydroxide (KOH) and ethylene-diamene pyrocatechol-water (EDP) are the most common so-called wet etchants. The etch rate is dependent on temperature and for this reason these etch solutions are usually warmed to around 100 °C to achieve a reasonable etch rate. It is essential that the temperature be kept stable during the etching process.

Techniques are available which allow the selective removal of silicon by gases or plasmas. These are referred to as *dry* etching materials. Dry etching is often preferred as it allows greater control than wet etching. The two mechanisms for dry etching are ion and reactive processes. With the former method etching is achieved through ion bombardment of the unwanted material. For reactive etching a radio frequency generated plasma reacts with the film to produce volatile compounds. Freon and chlorine gases are often used for etching silicon.

#### 5.2.3.2 *Micromachined Structures*
We have seen in previous chapters that a thin beam such as the cantilever or encastre-type provides a suitable basis for a transducing element. The extremely small beam geometries which are achievable with micromachined silicon techniques are particularly advantageous for devices like accelerometers where lightweight structures allow a wider range of applications. A simple cantilever structure can be etched out of silicon using a wet etch (EDP) technique. During processing it is possible to deposit piezoresistive elements at the fixed end of the beam thereby allowing the surface strain of the beam to be measured. Figure 5.3 shows a scanning electron micrograph of a series of cantilever structures etched out of s silicon wafer. The beams have different geometries and will therefore exhibit different resonant

frequencies. The examples in figure 5.3 show beams etched from silicon, but other materials are also used: polysilicon, silicon nitrode and silicon dioxide have all been successfully used as the base material for a number of micromachined sensors.

In addition to etching out structures it is also possible to create micromachined transducers by a silicon fusion bonding technique which allows the fabrication of structures like pressure diaphragms entirely from single crystal silicon. When this technique is combined with standard etching methods elaborate structures can be realized.

Recently there has been a great deal of interest in the area of micromachined actuators. The science, technology and design of moving micromechanical devices and mechanisms has been given the term *microdynamics* (Muller 1990). Both polysilicon and silicon nitride have been used as the mechanical structures. Owing to the extremely small physical size of such devices it has been possible to produce microactuators, like motors, which can be driven by electrostatic forces. There is an interesting reason why electrostatic actuators come into their own as the dimensions decrease. For large machines, electrostatic devices cannot compete with electromechanical devices because in the former case the field energy density ($\varepsilon E^2$)/2 is limited by breakdown, while in the latter case the field energy density ($\mu H^2$)/2 has no such theoretical limit. However, Paschen's law tell us that there is a minimum voltage below which breakdown cannot occur, so if the applied voltage

**Figure 5.3** A scanning electron micrograph of several micromachined silicon beams. (Courtesy Microelectronics Industrial Unit, University of Southampton.)

is kept below this value the energy density increases dramatically as dimensions decrease. In consequence the simpler configuration of electrostatic devices makes them very attractive in micro-engineered systems.

There is even the appearance of micromachined gear wheels, springs and sliding structures which several years ago, would have been thought of as nothing more than an academic pipe dream. To date, the number of applications has been limited by the modest load that micromotors are capable of driving. However, an enormous amount of research effort is being placed in this area and interesting and novel devices are continually being created. One particular device which has attracted attention is the so-called micropump (Bart *et al* 1990) which has been proposed as a way to administer drugs to a patient on a continuous basis rather than by other methods like self-injection.

## 5.3 THIN-FILM TECHNOLOGY

Another area of technology which has successfully been used in sensor design is that based on thin-films. First it is important to define what is meant by *thin-film*. Here the evidence in the numerous texts on the subject appears to be conflicting and sometimes confusing. Some authors define the term by referring to the thickness of the film itself while others characterize the term by the fabrication technique used to deposit the film. Now, while we do not want to get into a philosophical argument about our definition of terms, we do feel it is essential to distinguish clearly between a thin-film and a *thick-film* and for that reason we shall adopt the definition which differentiates between the two by the fabrication technology employed. However, typical thin-films have a thickness usually less than 1 μm and sometimes as thin as 1 nm. Since a single atom has a diameter somewhat less than 0.5 nm, thin-films are thus only a few atoms in depth. A number of different deposition techniques are available, and these can be divided into three main processes; evaporation, chemical vapour deposition (CVD), physical vapour deposition.

### 5.3.1 Evaporation

This process relies on the fact that when a solid is heated to a sufficiently high temperature in a vacuum it will evaporate and the evaporating atoms will travel in straight lines until they condense upon a suitable substrate. The process is illustrated in fig 5.4 which shows a bell jar which can be evacuated via a vacuum pump. Inside the bell jar is the substrate and the evaporation source which resides within a crucible. The evaporant is heated by one of several methods including direct resistance, electron beam, radiation and laser beam methods. It is usual to have

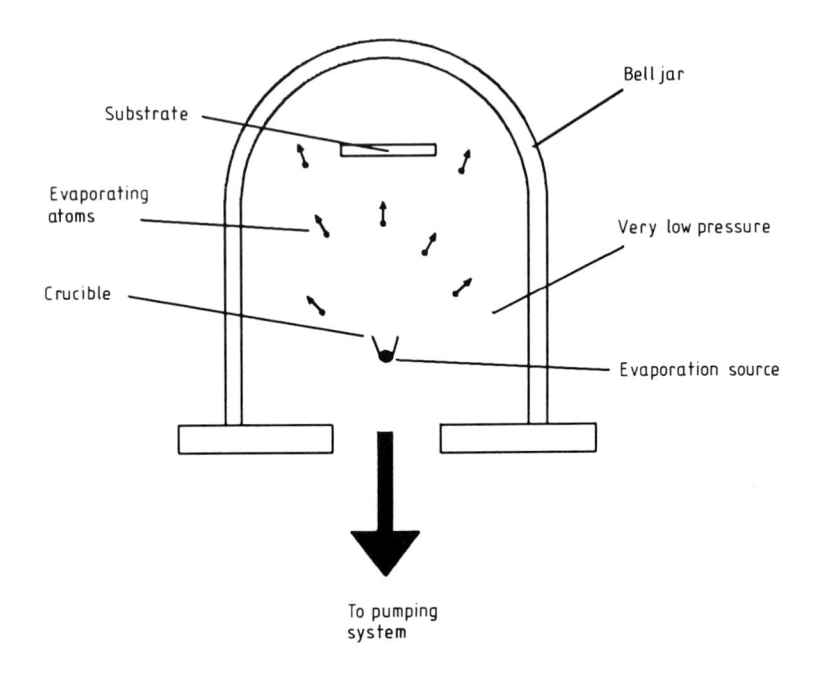

**Figure 5.4** A thin-film evaporation system.

control over the temperature of the substrate as this can be used to alter specific properties of the film.

The thickness of the film needs to be monitored while it is being deposited. A popular way of doing this is with a quartz crystal. The crystal is mounted inside the vacuum chamber close to the substrate. The quartz crystal itself is the frequency controlling element of an oscillator. When the evaporated material deposits on the crystal, the mass increases thereby reducing the natural frequency of oscillation. The change in frequency can therefore be calibrated against film thickness.

Metals like gold, aluminium and silver can all be deposited by vacuum deposition methods. Certain resistive films such as the alloy nickel-chromium, can also be fabricated using this technique.

### 5.3.2 Chemical vapour deposition

The two major types of chemical vapour deposition are the atmospheric pressure (CVD) and the low pressure (LPCVD). Materials like polysilicon, silicon dioxide and silicon nitride can be deposited by these two methods. Another type of vapour deposition is PECVD, plasma enhanced chemical vapour deposition, and this has the advantage of a lower deposition temperature than the previous two methods. Metal films are usually produced by the chemical oxidation of halide compounds.

### 5.3.3 Sputtering

If a surface is bombarded with high energy particles, atoms on the surface are ejected in all directions. This process is known as sputtering and is basically one of momentum transfer. The sputtered atoms can be condensed on a substrate in a vacuum in similar manner to evaporated films. Essentially there are two main types of sputtering techniques namely DC sputtering and radio frequency (RF) sputtering. DC sputtering, sometimes called glow discharge sputtering, tales place in a low pressure, inert gas atmosphere, usually argon. The glow discharge is formed when a high DC voltage is applied across two electrodes. The cathode is the source of the sputtered atoms and the substrate is usually attached to the anode. The high electric field causes the inert gas to become ionized and the positively charged ions are attracted towards the cathode. The ions bombard the cathode thereby causing surface atoms to be sputtered. A number of secondary electrons are produced at the cathode and these serve to maintain the glow discharge. Some of the sputtered atoms are intercepted by the anode and hence form a uniform thin film. A schematic diagram of a DC sputtering system is depicted in figure 5.5. The chemical composition of the deposited films can be modified by the addition of various reactive gases like oxygen to the inert gas. This is known as reactive sputtering and is often used to fabricate oxide layers.

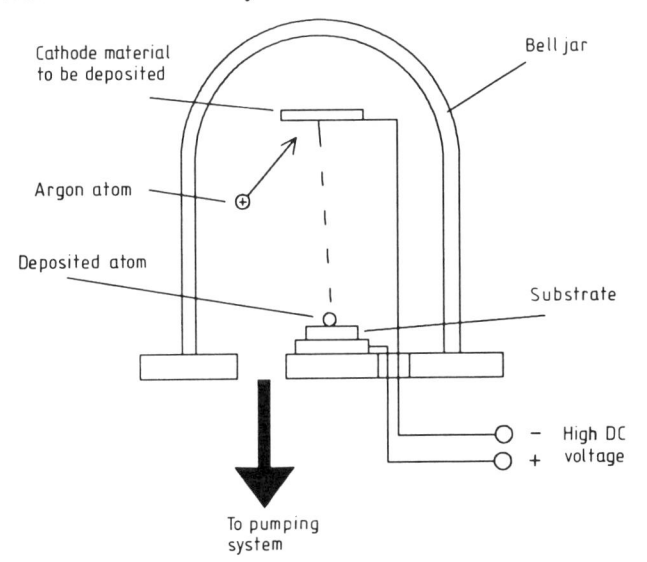

**Figure 5.5** DC sputtering system.

Insulating materials cannot be deposited using DC sputtering methods as the insulator would soon become charged and prevent further deposition. RF sputtering overcomes this problem, as the targets can be periodically charged and discharged. This process usually occurs at a high frequency of around 13.5 MHz. RF sputtering allows the deposition of conductors, insulators and resistors and is often employed to deposited resistive alloys such as nickel-chromium. Sputtered films, in general, exhibit better adhesion properties than evaporated films.

### 5.3.4 Langmuir-Blodgett films

Extremely thin films can be produced using the Langmuir-Blodgett (LB) methods. This technique relies upon the principle that wherever a gas or a liquid is in contact with another liquid or a solid surface there is a layer of molecules, often regularly oriented, at the interface. This interface region is called the chemical monolayer as its typical thickness is that of a single molecule. It has long been recognized that when oil is added to water the oil spreads out and forms a thin layer on the water surface. The water can act as a sub-phase and, if a suitable substrate is drawn vertically through the sub-phase, a monolayer of oil can be deposited.

Langmuir-Blodgett films have been applied to a number of sensor applications where high purity, thin layers are required. One particular area of interest which was highlighted by Langmuir himself is in the construction of thin, highly specific biological membranes allowing the study and control of biological reactions involved in the diagnosis of and treatment of disease. The membranes have been used in conjunction with ISFETs to produce highly selective enzyme sensors.

Many attempts have been made to optimize the efficiency of solar cells by incorporating a thin insulating layer between the photoconductor and the top transparent electrode. LB films have been employed in this area.

### 5.4 THICK-FILM TECHNOLOGY

Thick-films have successfully been employed in the realization of sensing devices in a number of ways (Brignell *et al* 1988), the three most significant being:
   1) The provision of the associated electronic circuits (i.e. power supply, signal conditioning etc.) in a compact form which is in a suitable form for incorporation into the sensor housing.
   2) The creation of support structures onto which sensing media can be subsequently deposited.
   3) The use of the thick-film material itself as a primary sensor.

Thick-film technology has been widely used in the hybrid microelectronics industry since the 1960s and is generally characterized by its compact, inexpensive and robust nature. As discussed earlier we shall distinguish between thick and thin-films largely by the fabrication processes used for their deposition, the significant difference being the screen printing method use for production of thick-films.

### 5.4.1 The thick-film production process

Thick-films are produced by the deposition of special inks onto an insulating substrate, the purpose of which is to produce a fired composite which controls the electrical conduction process. The most common substrate material is 96 % alumina which is usually purchased in the form of a ceramic wafer. The three main categories of ink are; conductors, resistors and dielectrics. Each of these contains three basic constituents, namely: the binder (in the form of a glass matrix), the organic carrier and the active material (metals, alloys, metal oxides or ceramics).

Conductor inks are required to have high conductivity and usually comprise a finely divided noble metal powder which accounts for around 80-90% of the total ink volume. Gold, silver, platinum and palladium are the commonly used active elements, and the particle size would be around 5 µm.

Resistor inks are similar in nature to conductor inks. The conducting phase is usually a precious metal oxide like ruthenium dioxide but the permanent binder and organic vehicle are essentially the same as those used in conductor inks. The final resistivity of the inks is dependent on the binder/conducting phase ratio. A wide range of ink *sheet* resistivities is available generally in the range of 100 mΩ / square up to 10 MΩ /square.

Dielectric inks have a variety of uses in hybrid circuit application and the ink formulation will vary according to the required application. For example, a cross-over dielectric ink will need to have a low dielectric constant to minimize capacitive coupling, whereas a thick-film capacitor dielectric material will need a high dielectric constant to maximize the capacitance per unit area. Encapsulation dielectric materials are also available and these are used for protecting resistors from environmental attack. Thick-film dielectric inks are also used for insulating metal substrates which are sometimes used in preference to alumina when there is a requirement for a strong, flexible substrate.

The process used for film deposition is based on one of the oldest forms of graphic art reproduction, namely screen printing. Figure 5.6 illustrates the basic screen printing process. The technique relies upon forcing the ink through the open areas of a mesh reinforced stencil onto the substrate. The screen is held at a distance of around 500 µm from the top surface of the substrate. As the squeegee traverses the length of the screen it deflects the screen fabric (usually stainless steel, nylon or polyester) and forces the ink through the open mesh areas of the screen. Behind the squeegee the screen peels away and leaves the required pattern printed on the substrate. The second stage of the process involves drying the ink to remove the organic carrier. This is usually achieved in an infra-red belt drier at a temperature of around 150 $^{o}$C. When the substrate emerges from the drier, the ink has partially

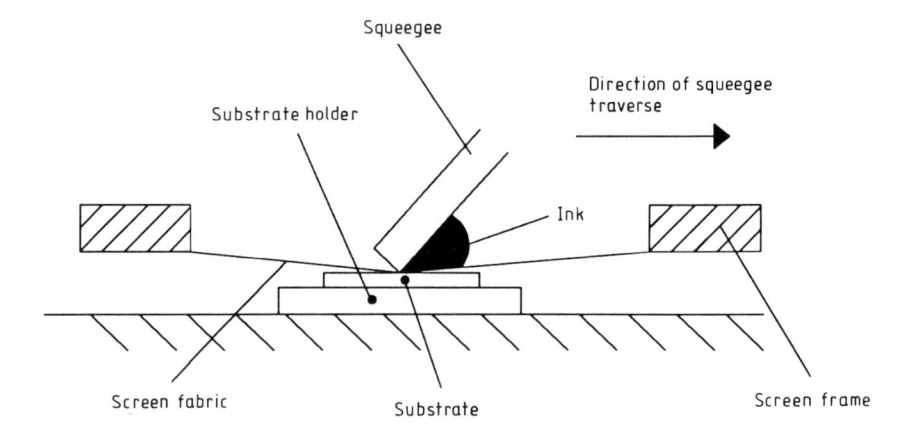

**Figure 5.6** Illustration of the basic screen printing process.

adhered to the base material and is immune to smudging. Some of the semi-permanent binder remains in the bulk of the film and serves to keep the particles fixed together to allow subsequent handling of the substrate.

The most crucial stage of the thick film process is the high temperature firing (or, more correctly, annealing) cycle. It is during this phase that the powders sinter to form a fired composite material. The peak temperature can be as high as 1000 °C. Firing is carried out in a continuous belt furnace usually in an atmosphere of clean air. Occasionally there is a requirement for an inert atmosphere like nitrogen but this is not very common. The temperature profile across the length of the furnace is required to have a ramp up to the peak, a constant plateau at the peak temperature for around ten minutes and ramp back down to ambient.

Some inks, particularly resistors, are sensitive to variations in the firing conditions and care is taken to ensure that suitable procedures are adopted to select the correct firing profile. The thick-film process is additive and layers are built up by repeating the basic print, dry and fire cycle. Some inks can be co-fired so that a print, dry, print, dry, fire sequence can be used.

### 5.4.2 Thick-film sensors

During the 1980s thick film sensors became an important category of solid state sensors because of their simplicity, robustness and low cost. Examples can be found of the use of thick-film sensors for all energy domains, namely mechanical, chemical, radiant, thermal and magnetic (Prudenziati and Morten 1986 and 1992, Brignell *et al* 1988)

The thick-film strain gauge is perhaps the best known example of a thick-film mechanical sensor. Its sensitivity to an applied strain is higher than that of a conventional metal foil gauge but lower than that of a semiconductor. However, thick-film strain gauges offer the advantage of lower temperature coefficients than those associated with silicon strain gauges. The technology was first employed in a commercial device in the early 1980s as the strain sensing element in a ceramic diaphragm pressure sensor for use in automotive vehicles (Catteneo et al 1980). Other applications have been realized by the successful deposition of thick-film strain gauges onto metal substrates (White 1988, Holford *et al* 1990, White and Brignell 1991). Low-cost accelerometers and load cells are two areas which have benefited from the use of insulated steel substrates.

Many commercially available thick-film pastes possess thermoresistive properties which are similar to that of the bulk material. Platinum ink is one obvious example where the temperature coefficient of resistance is both linear and repeatable thereby facilitating its use as a temperature sensor in the form of a platinum resistance thermometer (PRT). Thick-film PRTs offer the advantages of tight tolerances by means of trimming, ease of manufacture and a wide choice of shape and size (Reynolds and Norton 1985)

Thick-film technology lends itself to the construction of simultaneously fabricated arrays of sensing sites such as the one described by Brignell *et al* (1988). The construction of a chemiresistor was described in § 3.3.3.1. The interdigitated electrodes and heating element being easily fabricated in thick-film form leaving only the semiconductor oxide to be vapour deposited. However, there is evidence to show that materials like tin oxide and the phthalocyanines may also be deposited via a screen printing method. Thick-film electrodes also provide the basis of other chemical sensor structures for the detection of water partial pressure (humidity) and also in the fabrication of polarographic oxygen sensors for use in determining dissolved oxygen in sea water (Prudenziati and Morten 1986).

Thick-film sensors for radiant energy have been largely based on the photoconductive phenomena in screen printed films of cadmium sulphide and cadmium selenide. Photovoltaic conversion efficiencies of around 13 % have been measured. Studies have been made on the magnetoresistive properties of nickel-based thick films and measured values of around 0.6 % relative resistance change have been measured in a magnetic field of 2 kG. Applications include position and distance measurement (Prudenziati and Morten 1992).

## 5.5 OPTICAL FIBRE TECHNOLOGY

Optical fibres have been developed over a number of years largely to meet the needs of the communications field of electronics. With advantages such as a very wide bandwidth, immunity from electromagnetic interference and absence of a sparking hazard, it is no surprise that this area of work has attracted a great deal of interest. Aside from the traditional role of fibre optics for telecommunication applications they are also potentially attractive as a medium for sensor development.

Other advantages of optical fibres include factors like light weight, small physical size and chemically passive nature. The principle drawback of the technology is that of cost, not merely of the fibres themselves, but also of the detectors, couplers, optical sources (lasers, LEDs) and other components. This major disadvantage probably explains why, after much research and development effort, optical fibre sensors have made relatively little impact commercially. However, recent research has concentrated on areas where fibre optics are the only means of providing a suitable solution to a particular problem (Gambling 1990).

The fundamental principle behind the operation of optical fibre sensors is that the light along the fibre is modulated in some way. Common forms of modulation include intensity, phase, frequency and polarization. Systems in which the optical fibre itself plays a major role in modulating the transmitted light are referred to as **intrinsic.** In others the optical fibre simply guides the light to and from the modulating environment; this is termed an **extrinsic** system.

Silica glass is a very stable material which is strong and chemically durable. It is often used for information transmission in telecommunications. Many types of fibre optic sensor can be made from this material, thereby having the added advantage of being readily available. However, work has also been carried out on developing special sensor fibres with enhanced sensitivities to specific measurands (Birch *et al* 1982, Koo and Sigel 1982, Poole *et al* 1985)

### 5.5.1 Intensity Modulation

This is probably the simplest way of exploiting the sensing capabilities of fibre optics. Usually a reference is included as part of the measurement system in order to compensate for variations in the transmission properties of the fibre with time.

One example of an intrinsic sensor which relies on intensity modulation is the microbend sensor depicted in figure 5.7. The measurand can be one of a number of mechanical variables such as force, pressure, displacement etc. When this is applied to the sensor the amount of attenuation of the light at the bending points increases (in the case of a downward force) and the intensity of the light at the receiver decreases. The variation in intensity can be easily detected using a photodiode. This type of device has been used as a safety mat in an industrial environment to detect when someone is approaching or standing in a danger area.

### 5.5.2 Phase Modulation

The measurement of a phase change in an optical system can be achieved with an interferometer as shown in figure 5.8. Light from a coherent source, typically a

ENABLING TECHNOLOGIES

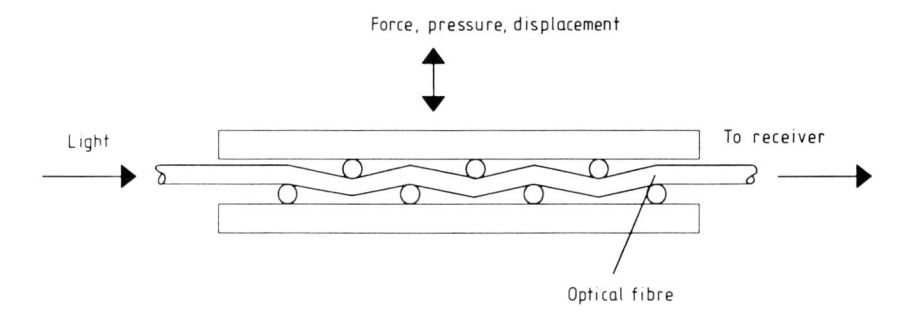

**Figure 5.7** A microbend fibre optic sensor.

laser, is aimed at a semi-silvered mirror which acts as a beam splitter and generates both the measuring and reference beams. The diagram illustrates a system capable of detecting the motion of a target prism which, when moved, will cause a phase shift between the measuring and reference beam.

An example of an optical fibre sensor which utilizes the principles of phase modulation is the optical fibre gyroscope shown in figure 5.9. Incident light from a source is separated by a beam splitter and is passed in opposite directions around a coil of optical fibre. The resulting phase shift $\Delta \varphi$ which occurs when the coil is rotated is given by

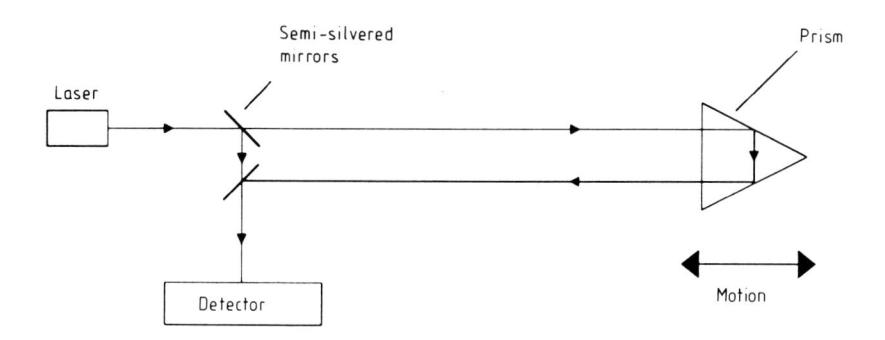

**Figure 5.8** A basic interferometer.

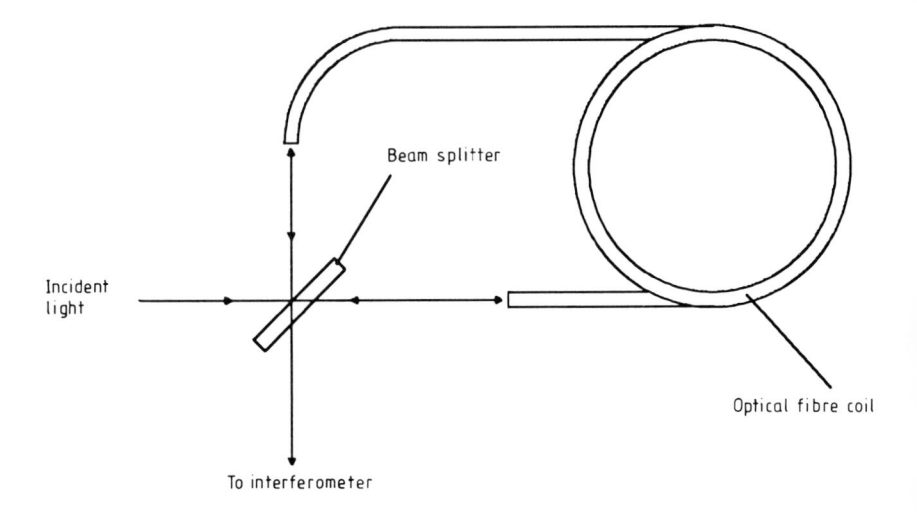

**Figure 5.9** Schematic of an optical fibre gyroscope.

$$\Delta \varphi = \frac{2\pi^2 d^2 N \omega}{\lambda c} \qquad (5.4)$$

where $d$ is the coil diameter, $N$ is the number of turns on the coil, $\omega$ is the angular velocity with which the coil turns, $\lambda$ and $c$ are the wavelength and velocity of the light respectively.

### 5.5.3 Frequency Modulation

A change in frequency of an energy signal relates to a shift in position on the electromagnetic spectrum. This spectral distribution is seen in the radiation emitted from a body when the temperature changes. An optical fibre pyrometer is one example of an extrinsic sensor, as the fibre is merely used to transmit the radiation to a measuring system.

### 5.5.4 Polarization Modulation

The main application of this technique is in a device referred to as an optical fibre current sensor. Operation depends on the Faraday magneto-optic effect which relates the angle of polarization of light in the fibre to the strength of magnetic field. If linearly polarized light propagates in the direction of a magnetic field of strength $H$, the plane of polarization is given by

$$\varphi = V \int HdL = VNI \tag{5.5}$$

where $\varphi$ is the angle of rotation, $V$ is the Verdet constant, $I$ is the current in the conductor, $N$ is the number of turns in the coil and $L$ is the path length. Hence if a coil of single mode optical fibre is wrapped around a conductor and the polarization rotation angle is measured then the electric current in the conductor can be easily found. Such devices can be used on high voltage transmission lines because glass itself is an insulator and it can be also be well separated from the conductor. Using this method it is possible to measure currents up to several kiloamperes with an accuracy of $\pm 0.1$ % (Laming *et al* 1988).

# 6

# Intelligent Sensor Concepts

## 6.0 INTRODUCTION

The rapidity of development in microelectronics has had a profound effect on the whole of instrumentation science, and it has blurred some of the conceptual boundaries which once seemed so firm. In the present context the boundary between sensors and instruments is particularly uncertain. Processes which were once confined to a large electronic instrument are now available within the housing of a compact sensor, and it is some of these processes which we discuss later in this chapter. An instrument in our context is a system which is designed primarily to act as a free standing device for performing a particular set of measurements; the provision of communications facilities is of secondary importance. A sensor is a system which is designed primarily to serve a host system and without its communication channel it cannot serve its purpose. Nevertheless, the structures and processes used within either device, be they hardware or software, are similar.

The range of disciplines which are brought together in intelligent sensor system design is considerable, and the designer of such systems has to become something of a polymath. This was one of the problems in the early days of computer-aided measurement (Brignell and Rhodes 1975) and there was some resistance from the backwoodsmen who practised the 'art' of measurement. Now, however, things are greatly changed, and modern electronic engineering courses cover virtually all the basic building blocks for the technology. Much of what is discussed in this chapter will be familiar to the recent graduate. It is in the form of a very brief survey, and the reader who finds the material unfamiliar would be advised to delve further into the literature.

## 6.1 ELEMENTS OF INTELLIGENT SENSORS

As we have seen (§ 2.0) the intelligent sensor is an example of a system, and in it we can identify a number of sub-systems whose functions are clearly distinguished from each other. The principal sub-systems within an intelligent sensor are:
  A primary sensing element.
  Excitation control.
  Amplification (possibly variable gain).
  Analogue filtering.
  Data conversion.
  Compensation.
  Digital information processing.
  Digital communications processing.

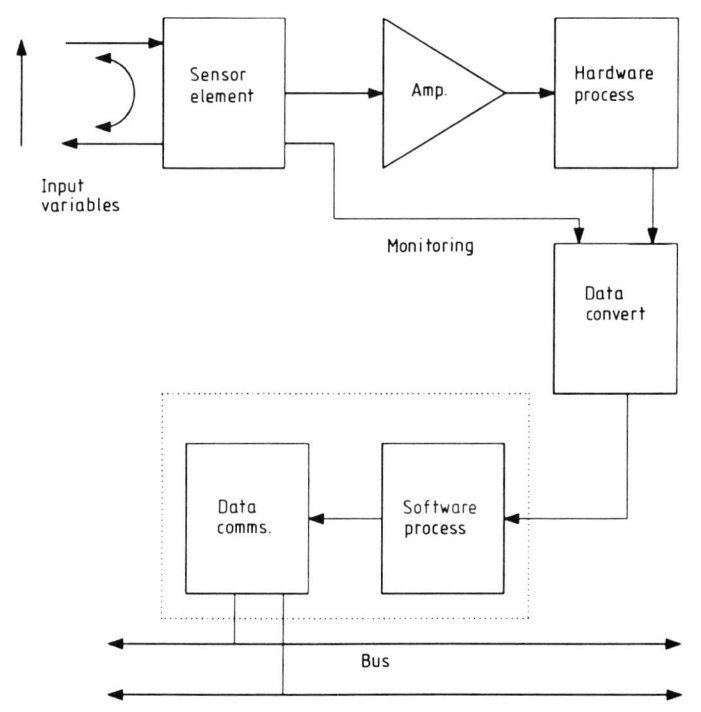

Figure 6.1 Elements of an intelligent sensor.

Figure 6.1 illustrates the way in which these sub-systems relate to each other. Some of the realizations of intelligent sensors, particularly the earlier ones, may incorporate only some of these elements.

**The primary sensing element**, which was our prime concern in § 3.3, has an obvious fundamental importance. It is more than simply the familiar traditional sensor incorporated into a more up-to-date system. Not only are new materials and mechanisms becoming available for exploitation, but some of those that have been long known yet discarded because of various difficulties of behaviour may now be reconsidered in the light of the presence of intelligence to cope with these difficulties.

**Excitation control** can take a variety of forms depending on the circumstances. Some sensors, such as the thermocouple, convert energy directly from one form to another without the need for additional excitation. Others may require fairly elaborate forms of supply. It may, for example, be alternating or pulsed for subsequent coherent or phase-sensitive detection. In some circumstances it may be necessary to provide extremely stable supplies to the sensing element, while in others it may be necessary for those supplies to form part of a control loop to maintain the operating condition of the element at some desired optimum. While this aspect may not be thought fundamental to intelligent sensors there is a largely unexplored range of possibilities for combining it with digital processing to produce novel instrumentation techniques.

**Amplification** of the electrical output of the primary sensing element is almost invariably a requirement. This can pose design problems where high gain is needed. Noise is a particular hazard, and a circumstance unique to the intelligent form of sensor is the presence of digital buses carrying signals with sharp transitions. For this reason circuit layout is a particularly important part of the design process.

**Analogue filtering** is required at minimum to obviate aliasing effects in the conversion stage (§ 2.2.4), but it is also attractive where digital filtering would take up too much of the real-time processing power available.

**Data conversion** is the stage of transition between the continuous real world and the discrete internal world of the digital processor. It is important to bear in mind that the process of analogue to digital conversion is a non-linear one (§ 4.5) and represents a potentially gross distortion of the incoming information. We have followed tradition in skirting the difficult implications of a non-linear system by dealing with it in terms of aliasing and quantization noise theory, which in general serve the purpose. It is important, however, for the intelligent sensor designer always to remember that this corruption is present, and in certain circumstances it can assume dominating importance. Such circumstances would include the case where the conversion process is part of a control loop or where some sort of auto-ranging, overt or covert, is built in to the operational program.

**Compensation** is dealt with elsewhere in this text (§ 3.2). Suffice it to say at this stage that it is central to the philosophy and reason for the existence of the intelligent sensor: also its needs may affect the basic design of the system, as exemplified by the presence of the monitoring line in figure 6.1.

**Information processing** is, of course, unique to the intelligent form of sensor. The range and variety of techniques which can be employed are far to broad to be treated here and many are discussed in detail later in this chapter, but it is worthwhile to summarize the general aims of using it in this application. There is some overlap between compensation and information processing, but there are also significant areas on independence.

An important aspect is the condensation of information, which is necessary to preserve the two most precious resources of the industrial measurement system, the information bus and the central processor. A prime example of data condensation occurs in the Doppler velocimeter in which a substantial quantity of information is reduced to a single number representing the velocity (§ 6.3.4.8). Sensor compensation will in general require the processing of incoming information, and in some circumstances will represent the major processing task. The intelligent sensor, to some degree, can be responsible for checking the integrity of its information; whether, for example, the range and behaviour of the incoming variables is physically reasonable.

As we have seen in § 1.3, it is a trite point, but nevertheless an important one, that the information processing stage cannot create information. It can, however, destroy information or introduce false information. This must be regarded as a major hazard in intelligent sensor design, as it is so easy to insert a process realized intuitively in software which may not be fully understood; the earlier use of the running mean (§ 2.2.5) is a case in point. Anyone involved in the real time programming of discrete information should have a working knowledge of discrete signal theory, certainly more than we have been able to touch on in § 2.2. The potential for the destruction of information is sufficiently important to bear re-iteration.

A final, but extremely important, element is communications processing. It is so important that it requires a processor of its own, though this may be realized as part of the main processor chip. The natural form of communication for the intelligent sensor processor is the multi-drop bus (Chapter 7), which can produce enormous cost savings over the traditional star topology network. A most important attribute of the intelligent sensor concept is addressability, which is of course essential to the multi-drop principle and a powerful aid to the logical organization of sensor systems operation, but it does introduce limitations. Addressability implies some form of polling of the devices, and though this may be prioritized in various ways, it does imply a constraint on the response time of the system to changes at any particular sensor site. A major contribution of intelligence is the integrity of communication. The transmission process can be protected by various forms of redundant coding, of which parity checking is the simplest example. In crucial applications information can be double checked by means of a high level handshake dialogue, in which the central processor asks for the information and then returns it to the sensor for confirmation. This deals with almost every possible fault except where the sensing element, though behaving apparently reasonably, is wrong. In such a case the only cure is the triplication of sensor elements, or in the extreme the triplication of intelligent sensors, as we shall discuss later.

## 6.2 STRUCTURES

### 6.2.1 Hardware structures

Hardware structures for intelligent sensors can reveal great variety (Brignell 1984). Obviously such structures are greatly affected by the enabling technologies employed. There is generally a need to mix technologies; an instrumentation amplifier, for example, poses different problems from a microprocessor, and to try to realize them in the same technology requires special and demanding circuit design techniques. Thus, while the single chip solution is something of a holy grail pursued by research laboratories, including that of the authors, it is something of a red herring in the design of effective intelligent sensors. For the time being a multi-chip solution is effective, which poses the choice of interconnection technology. Given the general need for miniaturization and mechanical robustness, this points to thick-film or silicon-on-silicon technology, which are rapidly succeeding the long established printed circuit technology (§ 5.1).

#### 6.2.1.1 *Minimal structure*
Consider the minimal hardware structure for an intelligent sensor. We have chosen, for the purposes of this text, to define an intelligent sensor as one containing digital processing, so a microprocessor is an essential element. Input amplification and data conversion are also necessary, which leads to the simple linear arrangement of figure 6.2. This provides a direct replacement for the dumb sensor, except that its defects are concealed internally and it behaves as though it were perfect. Note that a minimum requirement is the monitoring line at least for temperature which is an ever-present cross-sensitivity.

#### 6.2.1.2 *Analogue output*
Although digital systems are becoming more and more the norm there remains a requirement for analogue output. The intelligent sensor with analogue output effectively replaces the dumb sensor, but obviates its imperfections. It is, however, useful to discuss the implementation of analogue output, as it provides a platform for a discussion of a number of important intelligent sensor concepts.

The obvious way to implement analogue output is to provide a DAC, so that the output is available with a precision defined by that of the converter. There are, however, important variations which have some potential for improved performance. Figure 6.3 shows one of these. Instead of calculating the required output the digital processor calculates the difference between the actual amplified sensor signal and the ideal output (after corrections for non-linearity, drift etc.). This difference is output as a correction, which is added to the original signal in a summing amplifier. By judicious selection of the weighting applied by the summing resistors, this smaller corrections signal can be made to span the whole of the

# INTELLIGENT SENSOR CONCEPTS

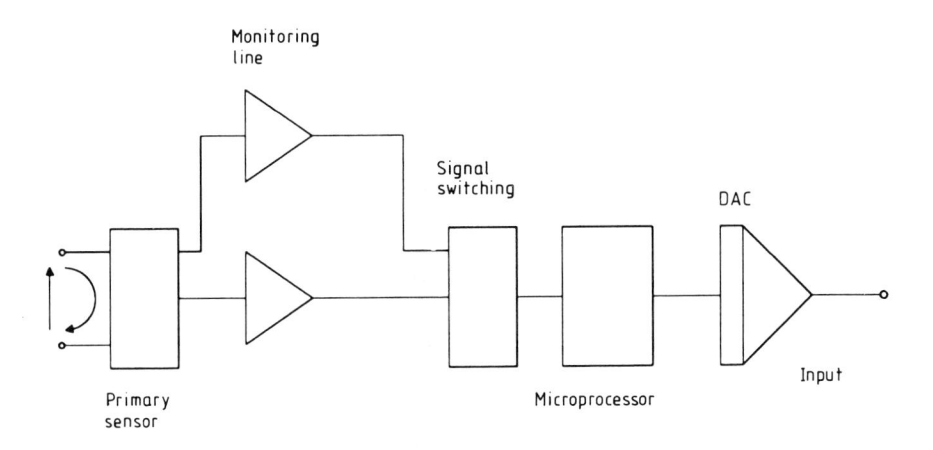

**Figure 6.2** A minimal intelligent sensor structure.

DAC output range. As a consequence the effective precision of the total output can be made much greater than the inherent precision of the DAC.

To take a simple example, imagine that the maximum anticipated error is 5% of full scale. A DAC of only eight bits could be used to span this 5%, and the effective precision at the summed output is one part in $2^8/0.05$, or better than 12 bits. Of course, as it stands this is not a satisfactory solution, since the errors in the summing resistors have to be taken into account, which is a useful point at which to emphasize the power of providing for an auto-calibration cycle. If we expand the system by providing a multi-way analogue switch at the input, which is under control of the processor, any errors associated with that switch will be common to all inputs. The extended system is illustrated in figure 6.4. This simple addition permits a variety of different calibration strategies. First, by switching in a reference voltage and then ground, the span in terms of voltage can be accurately assessed. Then, by switching in the output voltage a sweep of the whole range can be made to check for any sources of non-linearity, such as missing codes in the two data conversion stages. For temperature calibration the sensor is taken through its working range of temperatures. It is not necessary for the temperature to be known in any particular external units, but it is necessary for thermal equilibrium to be reached at a sufficient number of calibration points. Auto-calibration can be carried out on power-up, or it can be interleaved with carrying out the required functions of the sensor, provided the particular measurement strategy permits this.

It will be seen that the resulting sensor system is independent of errors produced by analogue component tolerances, and this exemplifies the intelligent sensor

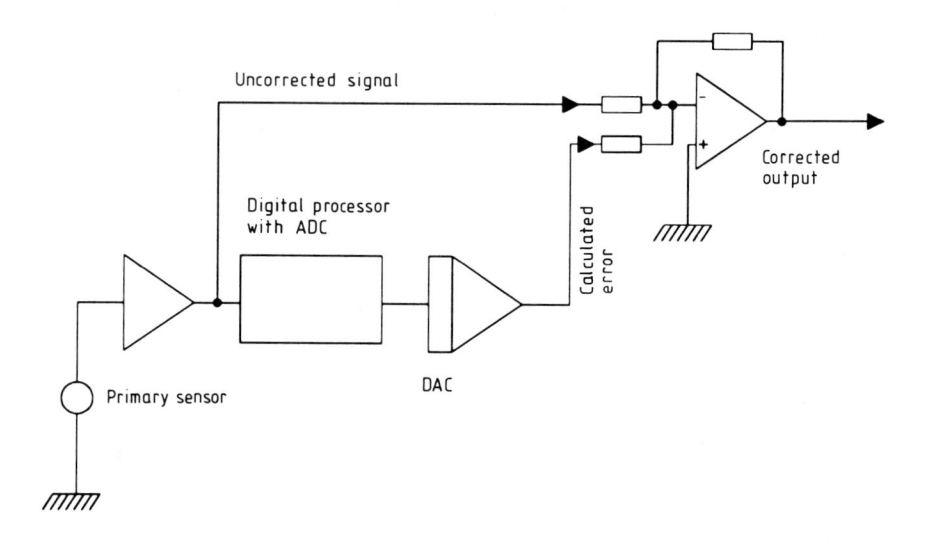

**Figure 6.3** An intelligent sensor with analogue output.

approach to accuracy. In the production stage there will generally also be a calibration cycle in which the target physical variable is swept through its range, and in certain applications it is possible to expose the sensor to calibrated signals in between operations.

### 6.2.1.3 *Self-check*
One of the most important characteristics of intelligent sensors is the provision of a self-check cycle. It was a major adverse criticism of the original concept that the extra complications would reduce reliability, and without self-check this complaint would be valid. In a system such as that shown in figure 6.4, the input signal switches allow a variety of test inputs to be applied. It should be noted that for the sake of clarity this figure has been somewhat simplified. It would, for example, be necessary to provide an attenuator so that the signals do not overload the amplifier. Also, without further switching elements (§ 8.5.1.3), we have created a complete feedback loop. This has two implications: the effect of the DAC voltage is diluted by an amount determined by the amplifier gain, and there are stability considerations.

A complete self-check cycle may be implemented as follows.

First, the input is switched to ground in order to check for any input offset drift. Small values of drift can be stored and applied as a digital correction, but large values of offset indicate a pathological condition which should be signalled via the communications processor as soon as the sensor is polled.

Second, a standard input derived from an internal reference voltage source is applied. This allows the gain to be checked and, if applied over a period, tests for intermittency.

Third, the input is switched to receive the DAC output via an attenuator, and a linear ramp is generated digitally. This simple test achieves a number of objectives. If the ramp is reproduced faithfully the linearity of the analogue components, the DAC and the ADC are all confirmed. Particular non-linearities that will be screened are missing bits in the DAC and ADC.

We should add here a remark on how the test procedure can be extended to differential amplifiers, as these are very common in intelligent sensors, because of the attractions of bridge configurations. An extra switch is required for the extra input, and the routine is as follows:

1) Apply zero volts to both inputs
2) Apply zero volts to one input and a reference voltage to the other
3) Reverse these connections
4) Apply reference voltage to both

These four stages allow for testing of the gain and the common mode rejection ratio. Ramp tests may be added if desired.

The completion of these tests ensures that all the electronic sub-systems are functioning correctly. The next stage is more difficult and very context dependent, and this is the testing of the primary sensor element. In some cases it is possible to arrange for known physical signals to be applied to the sensor, in which case a complete and proper calibration cycle can be carried out, but in the general case we have to assume that such a procedure would produce an unwarranted interference with the operation of the target system, and we have to make do with less satisfactory information.

Sometimes it is feasible to apply rather sophisticated methods. For example, it may be possible to apply a disturbing stimulus to the target system in the form of a pseudo-random sequence of a magnitude below the threshold that would interfere with operation. The response of the system and the sensor could then be recovered by a process of cross-correlation (§ 2.1.8).

Even without such elaborations it is possible to obtain some information to indicate whether the sensor is behaving correctly by, in effect, asking certain questions:

Is the output a reasonable value? That is to say, is it in range? Is it consistent with the prevailing conditions and plant history?

Is the rate of change of output reasonable? For example, a temperature sensor embedded in a thermal mass will have a constrained rate of response, and any more rapid changes would indicate some form of intermittency.

Is the output actually changing? In an active plant one would expect small changes to be occurring continuously. If they are not it is at least worth flagging a query to central control.

Is the output consistent with that of adjacent sensors? This question could be posed centrally, but it is also possible for the intelligent sensor to pick up the

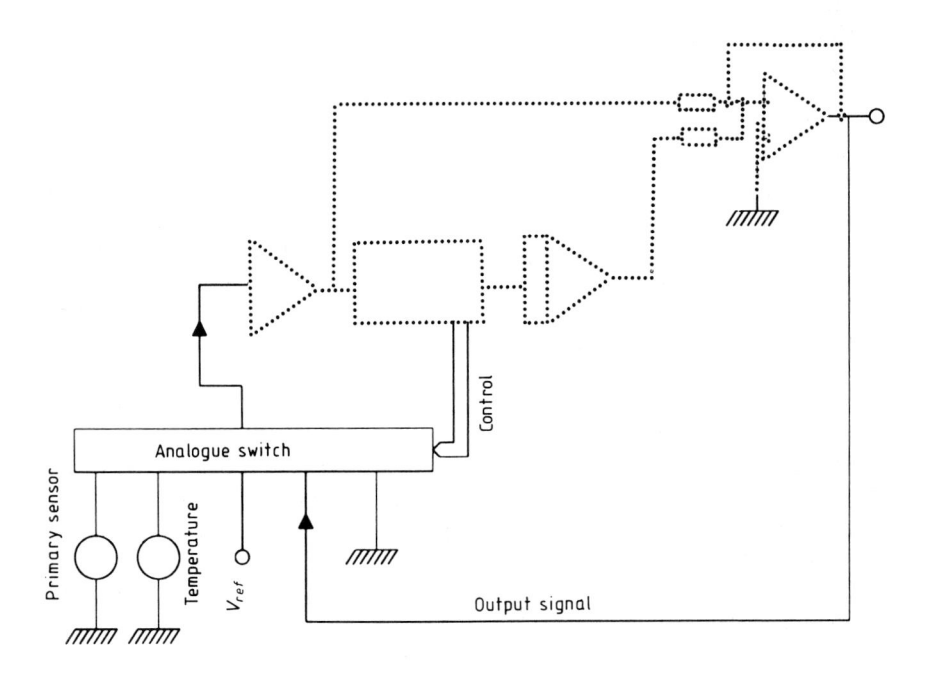

**Figure 6.4** The system of figure 6.3 extended to incorporate auto-calibration and self-check.

responses of its neighbours directly off the bus, thereby carrying out one of our prime requirements to relieve central control of unnecessary calculation.

There is, however, no escape from the fact that this part of the operation is extremely sensitive to context, and must be tackled case by case. Of course to obtain complete reliability one must resort to duplication or triplication. This option can be applied to the whole intelligent sensor, but as the electronic sub-systems are fully self-testing, a much cheaper option is to duplicate or triplicate the primary sensors, using signal switches to cycle between them. In the latter case the program can include a voting procedure, so that two correct primary sensors can outvote a discrepant one. It is important, however, to ensure that the pathological condition is signalled to control, because a second primary sensor failure would be fatal.

Often primary sensor faults are simple in nature and therefore easily detected, such as going open- or short-circuit, but there are many other faults that are slower to appear and more difficult to deal with. Examples are the accumulation of various forms of detritus, oxidation, fatigue and migration of materials. In such cases there is a stage at which it is not clear whether there is a fault or not, so it is important to establish within the high level communications protocol a means of signalling a possible incipient fault to prompt a human inspection before more damaging conditions are established.

Clearly this is an area where it is very difficult to generalize, but the above account establishes some of the principles that can be applied.

### 6.2.1.4 *General purpose structures*

In the early days of intelligent instrumentation it was particularly convenient to have available a system that was totally configurable by software, and it still has its advantages, particularly in the development phase. The main disadvantages are, first, that in designing such a system one is obliged to make decisions ( e.g. the speed-precision trade-off ) that will not always be appropriate and, second, in any application much of the system will be redundant. The latter point would not be important if a number of systems were manufactured, as the economies of scale would cancel out the waste.

Figure 6.5 shows diagramatically a system which was initiated as a project in the early 1980s by one of the authors (the so-called Janus project). It was realized as a circuit board and achieved a relatively brief existence as a commercial device (Brignell 1984). It is, however, very useful in the present context, as it illustrates many of the principles that had been developed up to that point and are major planks in the whole philosophy presented in this book.

The central component of this system is a digital controller of a number of analogue multiplexers. This controller is connected to the bus of the microprocessor and is mapped as a block of its memory, so that the system is reconfigured by writing control words to the block.

One set of analogue multiplexers provides signal selection to the differential input amplifier. This provides for sensor compensation by the sensor-within-a-sensor method or by the sensor array (§ 3.2, § 6.4.2). There is also provision for connection to an internally generated reference voltage and ground for self-check purposes (§ 6.2.1.3). Another multiplexer controls a resistor network to provide gain selection in the input amplifier. A 12-bit ADC gives data input into the microprocessor.

There is also provision for voltage output from an 8-bit DAC, which may be utilized as an external voltage for such applications as sensor excitation or as an internal offset to the input amplifier. Note the importance of offset provision before data conversion (Brignell 1986).

An essential adjunct to such a system is a software system which simplifies the interface problems for the user, and allows all the system settings to be made by means of simple high level commands (§ 6.2.2.5). A board such as that shown in figure 6.5 can be a very useful in intelligent systems development. Also present, but not shown, was a serial bus communications controller which allowed many such boards to be addressed on a single bus. Associated board level components were an intelligent bus controller which resided in a PC and an intelligent bus repeater, which allowed a network to be extended to kilometres in length.

152                    INTELLIGENT SENSOR SYSTEMS

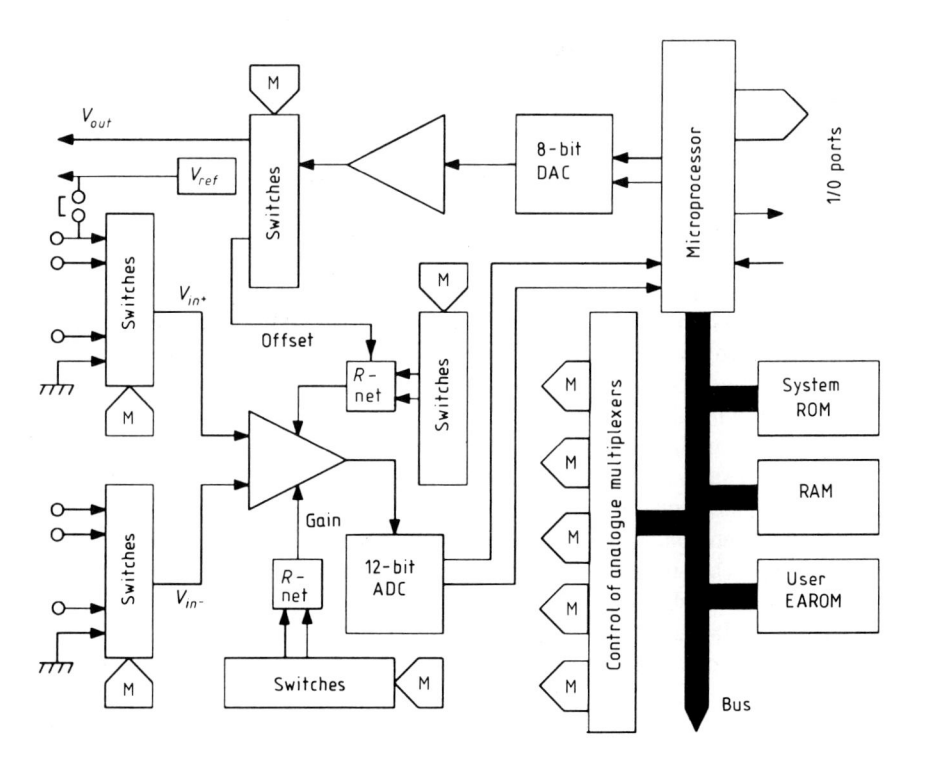

**Figure 6.5** Structure of a general purpose intelligence unit.

This example, though now somewhat obsolescent, merits study, as it adumbrates many of the principles that can be carried forward into the newer technologies.

### 6.2.2 Software structures

The art of good programming is a substantial discipline in its own right, and it would not be appropriate to delve into it in a short text such as this. It is, however, worthwhile to examine one or two structures that are peculiarly appropriate to intelligent sensor systems. In passing we might also make a remark about the choice of programming language. There are pressures in intelligent sensor design to work at the lowest level of programming, machine code, as this gives the highest speed and the most compact code. Where speed is not a special consideration, however, there are sound reasons for opting for a more portable language such as 'C'. The main argument in favour of this option is that it reduces the tendency to 're-invent the wheel'; since procedures can be programmed once and for all. It also reduces the necessity to learn a variety of low level languages and gives a common format which is universally understood.

### 6.2.2.1 *Look up tables*

One of the more powerful concepts that entered at the dawn of computing was the processing of arrays. The Look Up Table (LUT) is a simple example of array processing that is of enormous significance in intelligent sensors. The basic idea is that one or more input variables are used as pointers to values stored in an array, which are then used for further processing. The first and most prominent use of LUTs was in linearization (Brignell and Dorey 1983). Before the emergence of digital electronics non-linearity was a problem so overwhelming that it was almost universally avoided. The LUT changed all that, though the problem is now so reduced in importance that it is easy to forget that there are still pathological cases that are immune to correction (§ 3.2).

Another important application for our purposes is in the switching of sets of coefficients. If you require a number of digital filters, or cascaded sections of digital filters, it is not necessary to duplicate the code that implements a filter. It is merely necessary to change the base address of a pointer so that a new set of filter coefficients can be picked up from a different LUT, and use the same code with a different set of coefficients.

LUTs may also be of two or more dimensions. A very important application of multi-dimensional LUTs is in correction for cross-sensitivity. In this case one pointer will be derived from the uncorrected input variable, while the others will be derived from the interfering variables. By far the most important case is the two dimensional case in which the second variable is temperature, which is of universal concern as a cross-sensitivity.

Figure 6.6 shows diagrammatically the operation of a simple one-dimensional look up table. It is convenient to choose a size of table of $2^N$, where $N$ is an integer.

Assume that the input variable is derived from an ADC of $M$ bits precision. Then the top $N$ bits are masked off by a logical *AND* operation with the mask $(2^N - 1)2^{M-N}$. The masked value is shifted down $M$-$N$ places ( i.e. multiplied by $2^{N-M}$ ). This is now the incremental address that can be added to the base address (the location of the lowest entry in the LUT) to point to the desired value.

To illustrate the point with numerical values let us assume we have a table of size 32 ($N$=5), and the input ADC is of 8-bits precision. The input variable is masked off with the mask 11111000 and shifted right three places to give a five bit incremental address, which when added to the base address points to the required value.

Evidently we have thrown away three ($M$-$N$) bits of information. What we do next is a classical example of the trade-off between speed, precision and storage. We can ignore the loss and go for maximum speed, we can make $M$=$N$ and go for maximum precision at the expense of storage or we can use the bottom $M$-$N$ bits to provide a linear interpolation between the the selected entry and the next one up, and sacrifice some speed to gain precision. Indeed we can use higher degree interpolation formulae to gain precision and conserve storage at the expense of speed. The choice made depends on the demands of the particular application.

For reference in this case the linear interpolation formula reduces to

$$e = e_n + (e_{n+1} - e_n) 2^{N-M} r \tag{6.1}$$

where the corrected entry, $e$, is derived from the entry pointed to by the masked pointer, $e_n$, and $r$ is residue in the bottom $M$-$N$ bits. Higher order interpolation can be used for greater precision at the sacrifice of speed.

Figure 6.7 shows how a two dimensional LUT is arranged in the store and how it is thought of conceptually. If the two variables, say $x$ and $y$, are masked off to $M$ bits of precision, the new incremental address is formed from $2^M x + y$, so that the layout of the area of storage containing the LUT is as shown on the left-hand side of the figure. It is helpful, however, to conceive of the arrangement as a two dimensional one, as illustrated on the right of the figure.

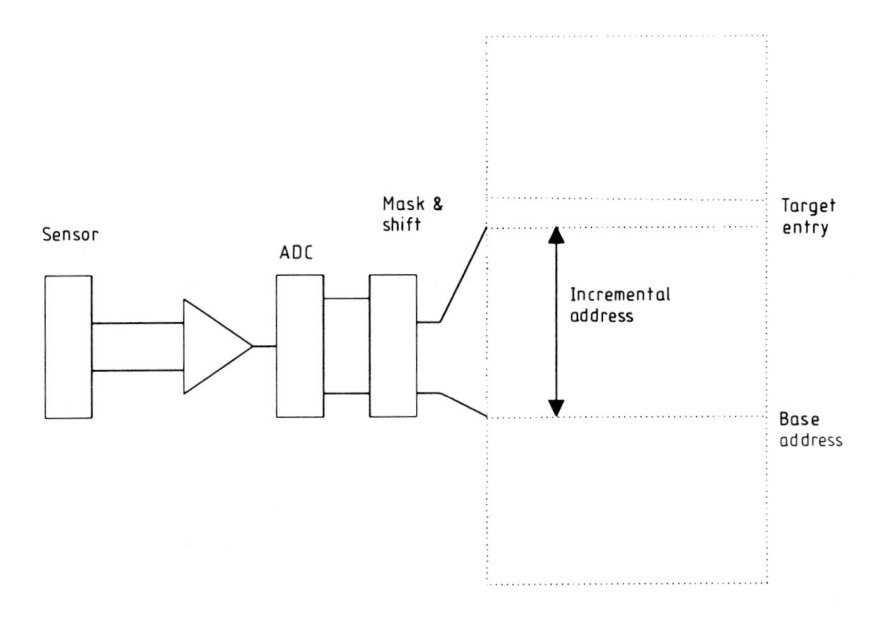

**Figure 6.6** Illustration of the operation of the look up table.

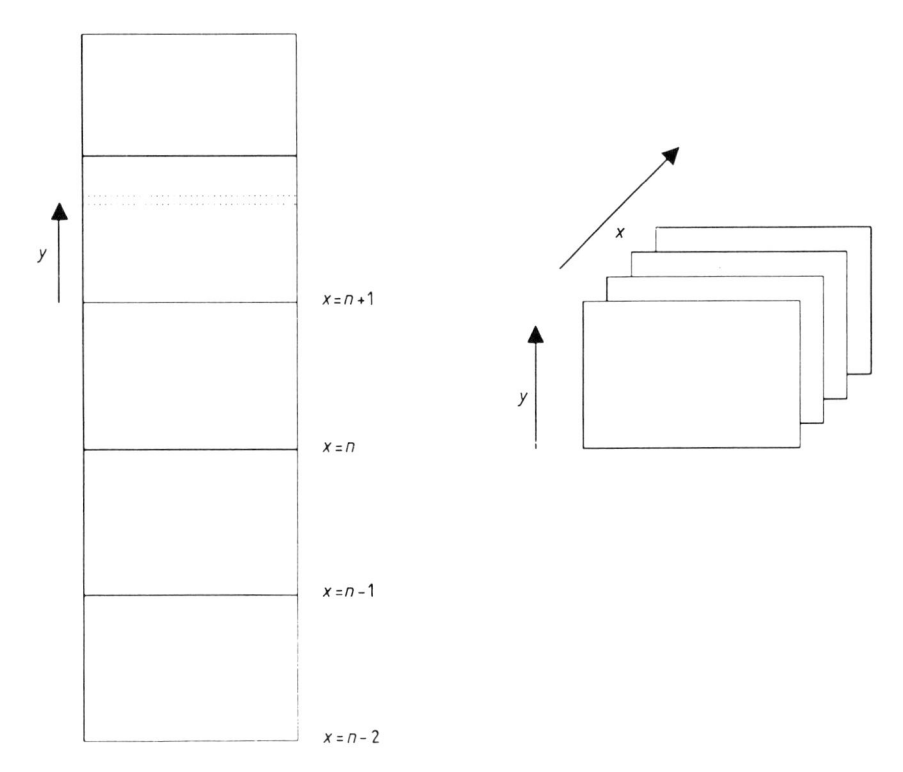

**Figure 6.7** The operation of a two dimensional look up table, as it is arranged in store and as it is conceived.

There is a variety of ways in which the entries can be loaded into the LUT They might be derived from a model, a common calibration curve for the family of sensors or (most preferably) by means of an individual calibration cycle.

*6.2.2.2 Cyclic buffers*
Another software structure important in intelligent instrumentation is the cyclic buffer. It can be implemented in a way similar to the LUT, in that it is based on masked pointers. These are incremented every time there is a read or write operation, and because the bottom $n$ bits are masked they return to zero every time they reach $2^n$. In this way, although the buffer is a linear array it behaves as though it were a circle, and the pointers behave like the hands of a clock (figure 6.8).

Cyclic buffers can have a number of useful applications. They are invaluable in linking two processes that are unsynchronized, such as a constant sampling rate and the random availability of communications access on a bus. We must always remember, however, that our simple law of information flow (§ 1.3) always applies, and the input and output demands of the buffer must align on average, or information will inevitably be destroyed.

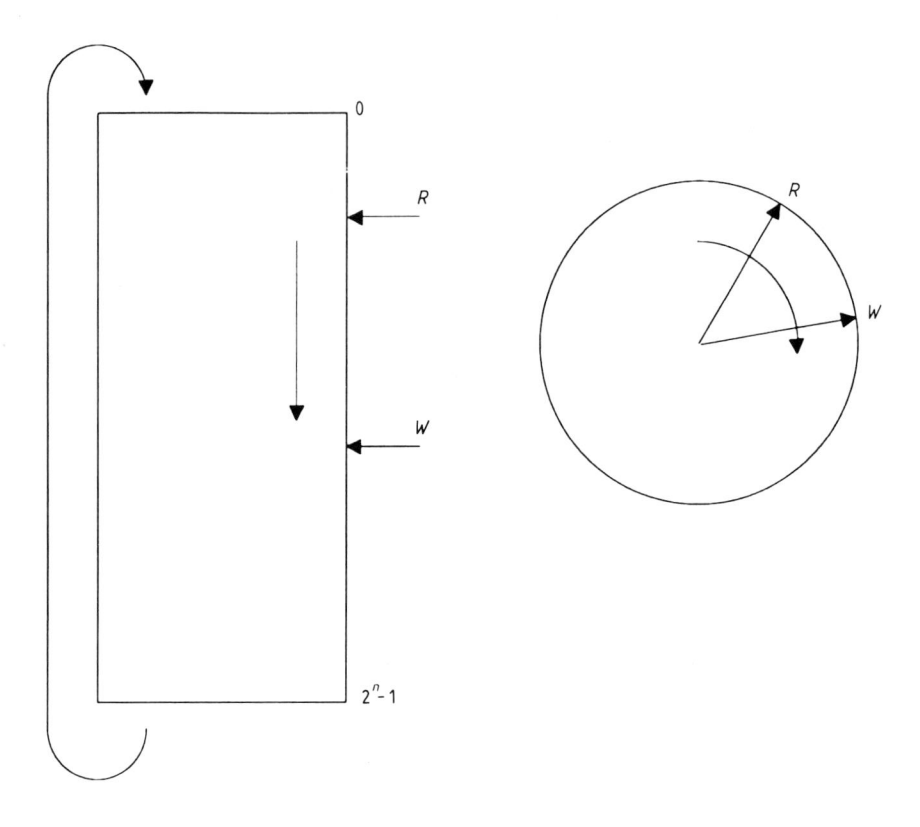

**Figure 6.8** The cyclic buffer, as it appears in memory and as it is conceived.

Another application of cyclic buffers is in the realization of digital filters and other processes requiring delay ( e.g. real time correlation ). Here the read pointers are linked rigidly to the write pointer to achieve fixed delays.

A third useful application area is in what might be called the software transient recorder. The transient recorder is a device that behaves like an oscilloscope except that it enables per- trigger information to be reproduced. The way it operates is that a signal is sampled continually, with each new datum erasing the one that came $2^n$ samples before. When a certain trigger condition occurs ( e.g. the signal reaches a given level ) the sampling process is stopped after, say, $k$ further samples have been obtained. There are then $k$ post trigger data and $2^n - k$ pre-trigger data. These can be displayed continually by reading them repeatedly to a screen, or transferred to a linear array, care being taken that the first datum is at the beginning of the array. The software transient recorder is invaluable in such applications as impulse or step testing, where it removes any need to synchronize with the stimulating signal.

### 6.2.2.3 *Signal processing structures*

When the numbers being processed are a time series, as is usually the case with sampled data from a sensor, then a powerful set of processes becomes available through the application of advances in linear algebra.

There are two main classes of signal process as far as we are concerned. These we will call **block** processes and **stream** processes. In block processing a finite number of samples is acquired, and the whole block of samples is processed once acquisition is complete. This is not normally a real time operation, though there are variations which make it effectively so. In stream processing the samples are acquired continuously and operated on as soon as they arrive. This is normally a real time process and the number of samples is effectively unbounded.

Block processing may be exemplified by the general linear algebraic relationship, where a new vector of variables, $y_j$, is obtained from the original set, $x_i$ by multiplication by a rectangular array of coefficients

$$y_j = \sum b_{ij} x_i \tag{6.2}$$

There is a powerful body of mathematics, known as linear algebra, which deals with such relationships. For us the most important example is the Discrete Fourier Transform (§ 6.3.4).

The most general form of the linear stream process is the recursion, which may be written

$$y_i = \sum_{1}^{n} b_k y_{i-k} + \sum_{0}^{n} a_k x_{i-k} \tag{6.3}$$

This is, in our terms, the recursive digital filter, which again, via the $z$-transform (§ 2.2.3), give us a powerful tool in manipulating signals. The power of the idea of recursion, in which a process can operate on its own outputs as well as the inputs, can hardly be overstated. We saw a simple example in the case of the running mean smoothing filter (§ 2.2.5), and the compaction obtained there is typical.

### 6.2.2.4 *Indirect software structures*

Most of our numerical methods are direct, or closed, structures, but it is easy to forget that the power of the digital processor enables us to use indirect, or open, methods. Many scientists and engineers seem to have a distaste for 'guesswork', because of their formal training in analytical methods, but often such methods yield significant solutions where none would be otherwise available (Brignell 1991).

The most familiar of such methods are the root finding iterations, such as Newton, but there is a highly developed set of tools which are in the nature of optimization. In order to facilitate an optimization technique there are three prior requirements;

1) A set of adjustable parameters or coefficients which fully define a process.

2) A measure of 'goodness'.
3) A means of making an educated guess at a better set of parameters.

Given these three factors we can use a method of trial and error to arrive at an optimum solution. Normally we also need a criterion for stopping the process, either sufficient accuracy or a limiting number of iterations, but in the sort of on-line process we find in intelligent systems it is possible to keep the process going on indefinitely, so that changes in the outside world can be tracked, and an operating point held at optimum.

The two basic classes of optimization process are gradient methods and gradient-free methods (Burley 1974). However, because the process of taking a gradient enhances high frequency noise (§ 4.5.4), we normally prefer the latter class in transducer applications. The method is best illustrated by an example, which we shall pursue in § 6.4.3.1.

### 6.2.2.5 *Software shells*

The idea of a software shell is now a familiar one, as the term is used with common computer operating systems. It is a very important concept with the sort of systems we are discussing here. The user of intelligent sensor systems is normally an

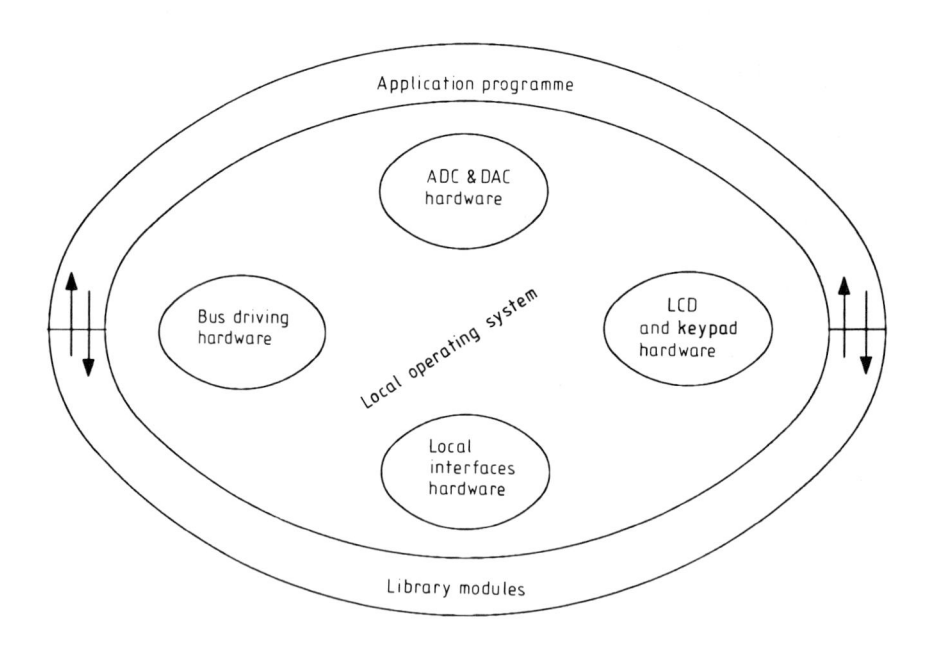

**Figure 6.9** Diagrammatic representation of a software shell which hides the complexities of local interfacing with hardware from the user.

instrumentation engineer, a Jack-of-all-trades who has not the time to delve into the intricacies of low level programming. It is therefore absolutely vital for the design of the intelligent sensor system to include the design of a software shell. This is not only important in protecting the user from the intricacies of internal operation, but like the shell of an egg it also protects the contents from being damaged by external events. Figure 6.9 shows diagramatically the software arrangements developed alongside the generalized hardware unit described in § 6.2.1.4.

## 6.3 PROCESSES AND PROCEDURES

### 6.3.0 Introduction

As we have seen in § 6.2, there are two important classes of digital signal process used in intelligent sensors, which we have called **stream** and **block** processes. An analogy is in watering the garden, where we might use a hose or a watering can. Stream processes work on a continual stream of sampled data, and the prime example is the digital filter. Block processes work on finite blocks of sampled data, and introduce the 'uncertainty principle' or 'window problem' (§ 6.3.4.3). Here the prime example is the Discrete Fourier Transform (DFT).

### 6.3.1 Digital filters

We have already met the idea of digital filters in Chapter 2 with the simple example of the running mean, where the $z$-transform theory enabled us to analyse its behaviour without difficulty.

Digital filters have a great deal in common with the more traditional continuous types, typical uses being to condition an input signal – e.g. high frequency noise removal (low-pass), trend removal (high-pass). Important differences do exist, however, and a digital filter may exhibit characteristics not possible with their analogue counterparts. The most fundamental difference is that a digital filter need **not** be physically realizable. For example, the filter need not operate in real- time, it may be designed to give zero phase distortion or to accept data in reverse order.

There may also be practical limitations imposed on analogue designs which can prevent their physical realization – e.g. component stability and size. This can be a particularly important consideration at low frequencies (< 1 Hz, Buttle *et al* 1968). Another problem, overcome by the digital approach, is the need to isolate the basic filter elements, which go to make up the complete system. Factors such as loading and impedance do not occur in digital signal processing.

### 6.3.1.1 *Filter Types*

A digital filter is a discrete-time system that performs some modification on an input sequence to produce an output sequence - i.e. it works on sampled data (Chapter 2). For this reason it is usual to describe its operation in terms of a suitable difference equation

$$y(nT) = \sum_{i=0}^{M} a_i x(nT-iT) - \sum_{k=1}^{N} b_k y(nT-kT) . \qquad (6.4)$$

The equation defines the $n^{th}$ output as a function of the $N$ previous outputs and $(M+1)$ most recent input values.

If all the $b_k$ terms are zero then the output is just a simple weighting of present and previous samples. This is termed a **non-recursive**, or **transversal, filter**. The non-recursive difference equation involving two delays (second-order system) is

$$y(nT) = a_0 x(nT) + a_1 x(nT-T) + a_2 x(nT-2T) . \qquad (6.5)$$

Note, each element of the expression involves the input value delayed by a multiple of the sampling period, $T$ i.e. in the $s$-domain multiplied by the unit delay operator, $e^{-sT}$ (see § 2.2). As we have seen a more powerful representation is the $z$-domain ($z = e^{sT}$) representation

$$Y(z) = \sum_{i=0}^{M} a_{-i} z^{-i} X(z) = a_0 X(z) + a_1 z^{-1} X(z) + a_2 z^{-2} X(z) . \qquad (6.6)$$

When both $a_k$ and $b_k$ terms are involved in the expression for $y(T)$, then the output is dependent on the previous input and output samples; analogous to the application of feedback in an analogue filter. The filter is then known as **recursive** and for the second-order case, the equation becomes

$$y(nT) = a_0 x(nT) + a_1 x(nT-T) + a_2 x(nT-2T) - b_1 y(nT-T) - b_2 y(nT-2T) .(6.7)$$

Recursive filters are generally favoured for digital implementations as the non-recursive types can often involve a large number of delay elements to realize a given characteristic. Furthermore, recursive designs, which have both poles and zeros, are the discrete counterparts of continuous-time filters for which a great deal of design information is available.

### 6.3.1.2 *Filter Design*

It is not appropriate to detail the design process here, but a brief summary might be useful. The usual method of design is to derive the transfer function in the

## INTELLIGENT SENSOR CONCEPTS

complex frequency domain, $H(s)$ ( $= Y(s)/X(s)$ ), and then to convert this, by one method or another to the z-domain, via the z-transform (§ 2.2). Before any design can commence, however, certain characteristics must be defined:
1) Class of filter, i.e. high-pass, low-pass, band-pass, etc.
2) Sampling frequency ($\omega_s$), i.e. the rate at which input is sampled.
3) Cut-off frequency ($\omega_c$), at which filter starts/stops to pass signal.
4) Rate of attenuation beyond cut-off.
5) Phase response.
6) Pass-band ripple.
7) Stop-band ripple.

Various filter types are available, all offering an approximation to the ideal rectangular response (figure 6.10) by trading one characteristic against another. The most common types are Butterworth, Chebychev and elliptic. The Butterworth filter gives the flattest amplitude response but with a poor cut-off performance. The Chebychev design improves the rate of attenuation at cut-off but allows ripple in the pass-band and the elliptic gives an even sharper cut-off but introduces ripple into both the pass and stop-bands. The designer must determine which of the characteristics is the most important for the particular application and select the most suitable filter type.

Once the filter type has been decided and the transfer function obtained in the frequency domain, it is necessary to transform this to the z-domain. Instead of a direct mapping between the s and z-planes, it is common practice to use a transform known as the **bilinear z-transform**. This maps the imaginary axis of the s-plane into a unit circle in the z-plane in such a way that the entire left-hand side of the s-plane maps into the interior of the unit circle. The right-hand half of the s-plane is mapped outside the unit circle. The technique involves the transformation of the continuous transfer function $H(s)$ in the s-plane into a new transfer function $H(s_1)$ in the $s_1$-plane that is periodic in $\omega$, with period $\omega = 2\pi/T$. The transform used to give this result is

$$s = \frac{2}{T} \tanh\left(\frac{s_1 T}{2}\right) \quad (6.8)$$

which, upon substitution of $z = e^{sT}$, yields

$$s = \frac{2}{T} \frac{z-1}{z+1}. \quad (6.9)$$

Several important advantages accrue from the use of the bilinear transformation, the most important being that aliasing errors (§ 2.2) possible with the direct z-transform methods, are removed. The major disadvantage of this method is that it produces a non-linear conversion of frequencies from the continuous system to frequencies in the discrete system. This is usually referred to as a *warping* of the frequency scale.

### 6.3.1.3 *Frequency Transformations*

Once designed, it is possible via simple transformations to convert a given filter to another type of the same class. For example, if a Butterworth low-pass filter is available in the form of its polynomial coefficients (or pole-zero locations) it may be converted to a high-pass, band-pass or low-pass filter of different cut-off frequency by the substitution of a given function of $z^{-1}$ for $z^{-1}$, i.e. $z^{-1} \to g(z^{-1})$ (Constantinides 1968, Papoulis 1980).

Care must be taken in the establishment of the transfer equalities to ensure the following conditions are met:

i) $g(z^{-1})$ must be a real rational function

ii) the values of the function $g(z^{-1})$ must correspond to those of $z^{-1}$ and no other values must be introduced

iii) the regions of stability must be preserved; the inside of the unit circle must map into the inside of the new domain and the outside to the new outside.

Despite its mathematical involvement, the technique is very powerful, for example, it can be shown that the transformation for a low to high-pass design is $z^{-1} \to -z^{-1}$ i.e. the signs associated with the odd powers of $z^{-1}$ ( i.e. the delays of odd order) in the original equation are changed.

### 6.3.1.4 *Practical Considerations*

The two main considerations for a digital processor implementation of a filter are:-

    i) Speed of operation (if filter is to operate in real time).

    ii) Processor wordlength.

The complexity of a filter will determine the amount of computation required on each input signal and will hence set a maximum sampling rate, above which the system breaks down. For real-time applications, this must be borne in mind when the input sampling rate is selected. The wordlength requirement in fixed point systems is a little more complicated. Firstly, sufficient bits must be available to specify the multiplier coefficients to a suitable accuracy since these determine the positioning of the poles and zeros on the $z$-plane (§ 2.2). If these are not positioned accurately then a different filter to the one intended will be implemented and in extreme cases, the poles may be moved outside the unit circle, and the filter becomes unstable. Sufficient bits must also be available for the representation of intermediate results – e.g. every multiplication of two $n$-bit numbers produces a result of $2n$ bits.

Provided the limitations of the processor in use are recognized, however, it is possible to implement a whole range of useful filters that are both stable and, more importantly, free from such errors as temperature drift or ageing of components. In general for the implementation of digital filters only two different sorts of basic block are required, a bilinear block

$$H(z) = \frac{a + bz^{-1}}{c + dz^{-1}} \qquad (6.10)$$

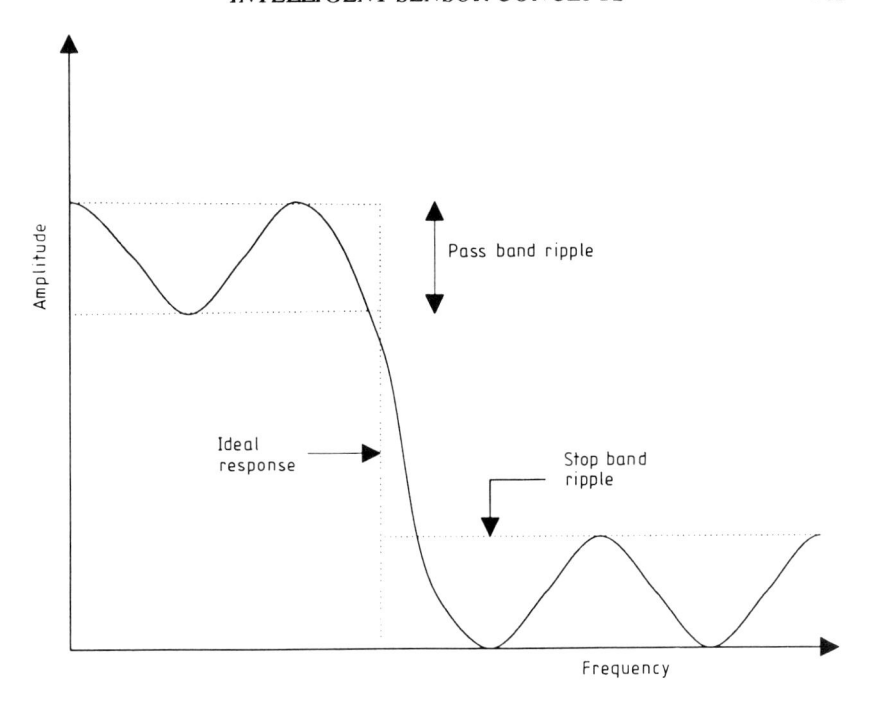

**Figure 6.10** A typical (elliptic) filter response compared with the ideal

which allows a real pole and a real zero to be implemented, and a biquadratic block,

$$H(z) = \frac{a + bz^{-1} + cz^{-2}}{d + ez^{-1} + fz^{-2}} \qquad (6.11)$$

which allows a pairs of complex poles and zeros to be implemented. Note that the code for these does not have to be repeated if blocks are cascaded, as new sets of coefficients can be switched in by changing an address pointer and the same code re-used.

### 6.3.2 Differentiation and integration

In § 2.2 we examined a process (the running mean) which is extremely easy to derive and implement, but which, if not fully understood in signal processing terms, may give misleading results. In contrast, differentiation is an example of a process which is difficult (indeed impossible) to implement and one which may be only approximated. A similar consideration applies to integration. Yet these two operations are commonly required by instrumentation engineers. Often this is because they are fundamental to the problem in hand (e.g. the information required is in the slope of, or the area under, the signal curve), but sometimes it is because thinking is restricted to the traditional continuous theory.

One of the reasons for the power of the Laplace transform is the fact that the operation of differentiation becomes simply multiplication by $s$ (§ 2.1.4). To determine how we can reproduce this operation by discrete processes we have to transform it to the $z$-domain, i.e. make the substitution from the defining relation $z = e^{sT}$, then

$$\frac{d}{dt} \to s \to \frac{1}{T} \log_e(z) \qquad (6.12)$$

There are three basic reasons why this result is not as simple as it looks. Firstly, the logarithm from a complex variable is infinitely valued. This is a direct manifestation of aliasing, since

$$\log_e(z) = \sigma T + j\omega T + 2\pi jk, \quad k = 0, \pm 1, \pm 2,\dots \qquad (6.13)$$

and as we have seen in § 2.2.4, it is impossible to distinguish between $\omega$ and $\omega + 2\pi k/T$. We shall set $k = 0$, thereby forbidding aliasing, and take the principle value. Secondly, our main interest is in the Fourier components $\omega$, which means we are primarily interested in the unit circle on the $z$-plane, i.e. the transform of the $j\omega$ axis. Thirdly, and most importantly, we are restricted to real powers of $z$ in our realization of the process (i.e. only delays or advances by multiples of $T$ in the time domain). Thus the best we can do is form a series, which to be practicable, must be truncated. It is beyond our present task to examine the methods for developing a series approximation for a function of a complex variable, so we shall give the result of applying the best known method, Taylor's series. This gives an expansion for $\log_e(z)$ in the region of $z = 1 + j\,0$ and we obtain

$$\log_e(z) = \sum_{n=1}^{\infty} \frac{(-1)^{n+1}}{n} (z-1)^n \;. \qquad (6.14)$$

So one form for the differential operator is

$$\frac{d}{dt} = \frac{1}{T} \sum_{n+1}^{\infty} \frac{(-1)^{n+1}}{n} (z-1)^n \;. \qquad (6.15)$$

This form will be found in textbooks on numerical analysis under the guise of the Newton Gregory forward difference formula. Thus the crudest approximation to differentiation can be made by selecting the first term only

$$\frac{d}{dt} = \frac{1}{T}(z-1) \qquad (6.16)$$

or in terms of samples $x_i$

$$\frac{dx}{dt_i} \approx \frac{1}{T}(x_{i+1} - x_i). \qquad (6.17)$$

More terms give better approximations.

Very often one finds that programmers apply the first order approximation for differentiation above indiscriminately, without consideration of the implications. We have noted that it is the first order of a series derived for the region of the unit circle near $z = 1 + j0$, which corresponds to $\omega = 0$, so we should expect it to be accurate only for low frequencies.

It is easy to show that the frequency response of the first order approximation (by substituting $z = e^{j\omega T}$ is given by

$$H(\omega) = \frac{1}{T}[\exp(j\omega T) - 1] = \frac{2j}{T} \exp\left(\frac{j\omega T}{2}\right) \sin\left(\frac{\omega T}{2}\right). \qquad (6.18)$$

So instead of the ideal characteristic for differentiation ($j\omega$) we have a sinusoidal form, which is a good approximation only at low frequency, see figure 6.11.

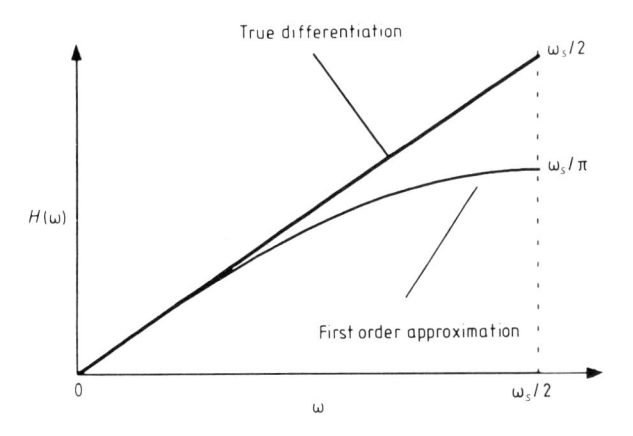

**Figure 6.11** The frequency response of true differentiation compared with that of the first order approximation.

Similar arguments apply to the problem of integration, which for similar reasons can also only be approximated. In general these will also involve formulae which are only valid for low frequencies. For fairly obvious reasons integration formulae are usually recursive and differentiation formulae non-recursive. The user of such digital processing methods must always be aware that they deviate from the theoretical ideal. It is thus important for the professional instrumentation engineer to think in discrete terms, and be ready to calculate the actual response in the frequency domain of any procedure he uses. There are many different formulae in the literature for differentiation and integration (Johnson and Riess 1977, LaFara 1973) of greater or lesser complexity. In certain circumstances the simple first order formula above is preferable to true differentiation as it amplifies high frequencies less. This emerged in our discussion of quantization, and in figure 4.20 true differentiation would have yielded infinite excursions of the function.

While differentiation emphasizes high frequency noise, integration emphasizes low frequency noise, particularly drift and $1/f$ noise (figure 3.4). One application of integration, provided the target signal is appropriately defined, is in drift tracking and removal by subtraction.

### 6.3.3 Smoothing

In the classical textbooks of numerical analysis many formulae will be found for the smoothing of functions (LaFara 1973), some of them of great complexity. In our terms smoothing is a low-pass filtering function, whether it is carried out by direct filtering or Discrete Fourier Transform methods. For this reason we will not dwell here on specific smoothing formulae, but merely remark that it is often a requirement for noisy signal sources.

### 6.3.4 Discrete transformation

Historically the idea of the frequency domain first entered the realms of applied mathematics in discrete form. This was when Fourier established that any function which satisfied a certain set of conditions known as Dirichlet's conditions, and which was either periodic or defined in a restricted region so that it could be regarded as periodic outside that region, could be expanded as a series of sine waves. Since the sine waves have known simple properties for the operations of differentiation and integration this leads to a means of solving a wide range of the equations which describe physical systems, particularly where arbitrary functions are used to describe the initial situation. By this means many of the practical problems of engineering became soluble for the first time (e.g. heat flow, vibration, electromagnetic fields and optical calculations). The power of this idea of expanding a function in terms of a fundamental and a series of higher harmonics arises from the fact that in linear systems the superposition principle applies, so that each harmonic may be treated separately and the results added to produce the general solution. The Fourier series may be written

$$f(\theta) = \frac{a_0}{2} + \sum_{n=1}^{\infty}(a_n \cos n\theta + b_n \sin n\theta) \qquad (6.19)$$

and the coefficients may be evaluated by means of the integrals

$$a_n = \frac{1}{\pi}\int_{-\pi}^{+\pi} f(\theta) \cos n\theta \, d\theta \qquad b_n = \frac{1}{\pi}\int_{-\pi}^{+\pi} f(\theta) \sin n\theta \, d\theta \ . \qquad (6.20)$$

A more generalized representation of the series can be obtained by using the well-known (*De Moivre*) relationship between the complex exponential function and the sine and cosine functions. Thus introducing the idea of a time series by substitution of $\theta = \omega t$ we have

$$f(t) = \sum_{n=-\infty}^{+\infty} c_n \exp(jn\omega t) \qquad (6.21)$$

where the generalized complex coefficients can be obtained by the relation

$$c_n = \frac{\omega_1}{2\pi}\int_{-\pi/\omega_1}^{\pi/\omega_1} f(t) \exp(jn\omega t) \, dt \ . \qquad (6.22)$$

The mathematical justification for the simplicity of this expansion is based on the property of orthogonality of the sinusoidal functions, which is very closely analogous with the idea of orthogonality in vector space and the representation of an arbitrary vector in the form of orthogonal components.

In the days before the advent of digital computers, it was found that the practical utility of the series did not measure up to its theoretical promise, particularly in situations where convergence was slow, because of the sheer amount of computation involved. The idea of the Fourier integral was developed to assist in these situations and particularly for the case where the fundamental interval extends to infinity.

The development of powerful methods of continuous linear electronics made the Fourier integral even more important, and electronic engineers began readily to think in terms of the frequency domain as a powerful alternative to the time domain in the representation of signals, as we have done in § 2.1.4. The Fourier transform may be thought of as a continuous development of the Fourier series; so the above summation becomes

$$f(t) = \frac{i}{2\pi} \int_{-\infty}^{+\infty} F(\omega) \, exp \, (j\omega t) \, d\omega, \qquad F(\omega) = \int_{-\infty}^{+\infty} f(t) \, exp \, (-j\omega t) \, dt \, . \qquad (6.23)$$

The Fourier transform became a very important factor in the theory of engineering and science, but, because of the difficulty of evaluating an arbitrary integral, its practical utility was very limited. This situation was completely changed by the development of methods of digital computation, and much of the early work of digital computers was devoted to the evaluation of Fourier integrals. However, because of the necessity for the computer to work in a discrete mode, the integral had to be treated in a discrete way, which was effectively a return to the Fourier series method.

It was soon found that difficulties arose in practice for the evaluation of Fourier transforms involving a large number of ordinates, because the implication of the above formula for the Fourier coefficients is that the number of calculations increases as the square of the number of ordinates. Not only did this impose limitation through the sheer length of the time required for the evaluation of practical problems, but there was also the problem of the decrease in accuracy which occurs through such a long chain of calculations. The discovery of the fast Fourier transform was therefore a very significant event.

### 6.3.4.1 *Properties of the Discrete Fourier Transform*
The discrete transform, as may be seen by its relationship with Fourier series, possesses the property of periodicity. This must be treated with great care, as must the sampling theorem to which this property is closely related. The implications of this are, first, that the transform can only be applied to a finite interval and, second, that in the interaction of functions account must be taken of this periodicity or circularity. Figure 6.12 shows the relationship between the idea of the circularity of a function and its periodic extension.

It will be noted that there is a similarity between the circular basis of the Fourier transform and the importance of the unit circle in the $z$-transform. This is no accident, and a Discrete Fourier Transform (DFT) may be seen as the evaluation of the $z$-transform on the unit circle (just as the continuous Fourier transform is the evaluation of the Laplace transform along the $j\omega$ axis).

Like any other mathematical manipulation, the Fourier transform cannot create information, and so if $N$ independent values are put into the transformation no more than $N$ independent values can be produced by it. Hence, if we provide the transformation with $N$ samples of a function in time we can obtain precisely $N$ samples in terms of frequency. In fact the situation is slightly more complicated than this as the set of real samples in time is in effect a set of complex numbers with zero imaginary parts, so $N$ real numbers will produce $N/2$ independent

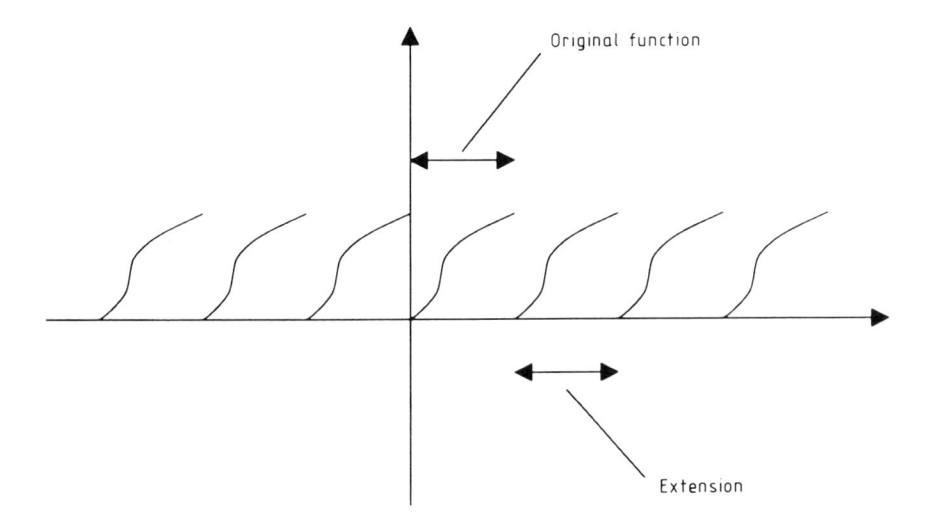

**Figure 6.12** The circularity of a function viewed as a periodic extension.

complex numbers on transformation, another manifestation of aliasing. Almost invariably it only makes sense to treat these samples as equally spaced both in time and frequency. Figure 6.13 shows the relationship between the $z$-transform and the DFT for 8 ordinates.

Much of the power of the Fourier transform is embodied in the convolution theorem(§ 2.1.7). This states briefly that the Fourier transform of the operation of convolution is the much simpler operation of multiplication. As we shall see this property can be exploited very effectively in the application of the DFT. However, it is most important to note that the form of convolution associated with the discrete Fourier transform is a circular or periodic convolution, which without care can produce serious errors due to the unexpected interaction between the beginning and the end of a block of data to be transformed.

This observation leads to a further important restriction on the use of discrete transformation, which arises from the fact that in nature events are very rarely broken up into complete discrete blocks of time, and the arbitrary division for the purposes of transformation can be a serious corruption. We shall examine this under the heading of the Window Problem.

### 6.3.4.2 *The Fast Fourier Transform*

The history of the fast Fourier transform is a fascinating subject (see the account by Cooley *et al* in Rabiner and Rader (1972)). When the method was described by

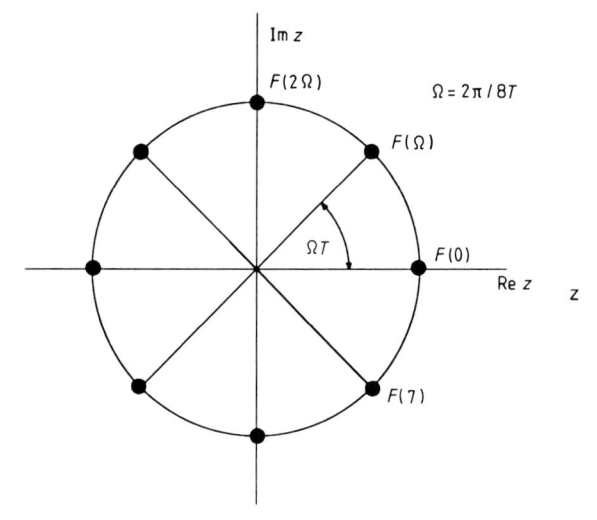

**Figure 6.13** The DFT as an evaluation of the function on the unit circle in the $z$-domain.

Cooley and Tukey in 1965 it was widely accepted as being original, but it subsequently emerged that this was a rediscovery, and the fundamental method had been used many years before. However, until this rediscovery, it was widely accepted that the $N$-point discrete Fourier transform required $N^2$ basic operations for its calculation. This had been proving a very serious bottleneck in a number of fields, for example, X-ray scattering.

In computational terms the Fourier transform is a relationship between two finite sequences of numbers $A_r$ and $X_k$, and this relationship is expressed by the following equation

$$A_r = \sum_{k=0}^{N-1} (X_k) W^{rk} \qquad r = 0, 1, \ldots, N-1 \qquad (6.24)$$

where

$$W = exp(-2\pi j/N) . \qquad (6.25)$$

Furthermore, it is not difficult to show that the inverse of this equation is

$$X_m = \frac{1}{N} \sum_{r=0}^{N-1} A_r W^{-mr} \qquad (6.26)$$

## INTELLIGENT SENSOR CONCEPTS

The factor of $N^2$ arises from the fact that each of the $N$ ordinates is itself generated by $N$ terms of the summation. However, the fact that $W$ (the $N^{th}$ root of unity) has circular properties means that there is a great deal of redundancy in these equations.

It is beyond the present treatment to examine in detail the methods by which these redundancies may be exploited to reduce the amount of computation, but as an example we may refer briefly to one of them. This is the method of decimation in time. The basic idea is that the sequence of $N$ ordinates may be treated as two separate subsequences; those of odd number and those of even number. Let us call these

$$g_m = X_{2m} \qquad m = 0,1,2,...,\frac{N}{2}-1 \quad . \qquad (6.27)$$
$$h_m = X_{2m+1}$$

The Fourier transforms of these sequences are also sequences of $N/2$ points which may be written

$$G_k = \sum_{m=0}^{\frac{N}{2}-1} g_m \left(W^2\right)^{mk}$$

$$(6.28)$$

$$H_k = \sum_{m=0}^{\frac{N}{2}-1} h_m \left(W^2\right)^{mk} \quad .$$

Now the DFT of the entire sequence could be written in terms of these coefficients, thus

$$A_k = \sum_{m=0}^{\frac{N}{2}-1} \left[ g_m W^{2mk} + h_m W^{(2m+1)k} \right] \qquad (6.29)$$

and we can show that

$$A_k = G_k + W^k H_k \quad . \qquad (6.30)$$

This relationship has very important computational implications, for even if we use the direct method, the term on the left hand side would have required $N^2$ operations (an operation being a complex multiplication and addition), while the terms on the right require $(N/2)^2$ operations each and combining them gives an additional $N$ operations, the total being $N + N^2/2$ operations.

If $N$ is the power of 2, it is easy to see that the above operation of decimation in time can be carried out continuously until we are left with sequences of only 2 points each. The result is that the discrete transform now only requires $N \log_2 N$ operations. In practice, a few further computational tricks are required to allow this to be performed conveniently (Gold and Rader 1969) but this is the essence of the fast Fourier transform methods. The method brings a great deal of power not only to a discrete transform itself, but also to filtering convolution, autocorrelation, two dimensional transforms (pattern analysis) and two dimensional filtering.

### 6.3.4.3 *The uncertainty principle*
A very important theorem in Fourier transform theory relates a change of scale in time to the corresponding change of scale in frequency. Thus

$$f(at) \rightarrow \frac{1}{|a|} F\left(\frac{\omega}{a}\right) \tag{6.31}$$

From this we can see that stretching out a function in time compresses its frequency domain version and vice versa. We may loosely verbalize this process by saying 'The more we can locate an event in time the less we can locate in frequency'. This is one of the most important tenets of modern physics through the application of wave mechanics to matter. It does, however, embody a very important restriction on the manipulation of the time and frequency domains. In the extreme, an event which is exactly located in one domain (the impulse function) is represented in the other domain by an infinite spread (a constant).

This observation exposes one of the great difficulties in the application of Fourier transform theory to the processing of real signals. This is because in the real world signals tend not to start and stop conveniently so that we can sample them and store them in a finite block of computer memory. It follows that the result of chopping off a length of signal for computer processing is to spread its frequency domain version artificially. This can be a serious distortion.

### 6.3.4.4 *Window problem*
The practical implication of this last point gives us what is known as the Window Problem. This we can illustrate by examining one of the most important Fourier transform pairs in figure 6.14. This is the block function and the *sinc* function. In chopping off the block of signal we are effectively multiplying by the block function, which means that in the frequency domain we are convolving by the oscillatory *sinc* function. Thus, not only are we spreading out the function in the frequency domain as implied by the uncertainty principle, but we are also adding an oscillatory element. Hence, we find that if we perform a discrete Fourier transform on a finite block from a real signal, we introduce irrelevant minor lobes in the spectrum. To overcome this problem of the rectangular 'window', it has become the practice to use other sorts of window, i.e. pre-shaping functions to treat the signals before or after transformation. What these do in effect is to replace the

# INTELLIGENT SENSOR CONCEPTS

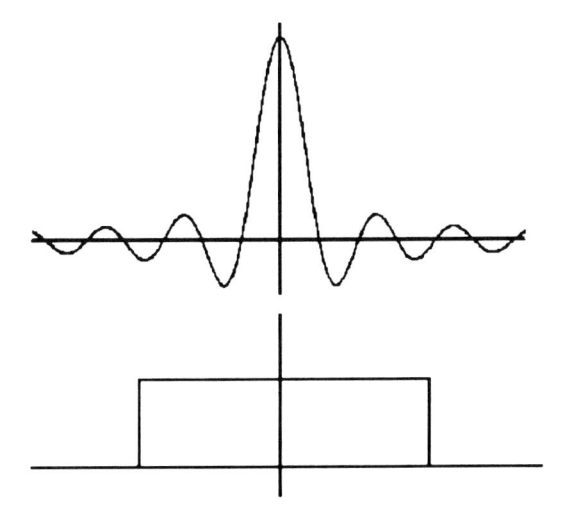

**Figure 6.14** The most important Fourier transform pair – the *sinc* and block functions.

sharp edges of the block function, which have high frequency implications, by a smoother transition, e.g. part of a cosine curve. One of the results of this is that it is quite possible, even probable, for two different workers to arrive at different estimates for the spectrum of a sampled signal. Thus, the transformation (or indeed any other form of processing) of a block of signal extracted from a continuing source must be treated with a great deal of circumspection (Oppenheim *et al* 1983).

Nevertheless, provided we always bear in mind the two limitations of the Sampling Theorem and the Window Problem, the discrete transformation yields extremely powerful methods of the analysis of signals, including the whole range of methods which come under the heading Spectral Analysis. We shall finish by examining briefly one important example (§ 6.3.4.8).

### 6.3.4.5 *Deconvolution*

We have seen (§ 2.7) that in the time domain the relationship between the input signal, the output signal and the system response is given by the convolution integral. This is fairly complicated and can be difficult to work out in practice, but the inverse problem of deconvolution is even more difficult. This means that to answer the question 'Given the output of a known system, what was the input?' We have to solve an integral equation.

The equivalent problem in practical electronics is the one of equalization, in which we need to remove the effect of an unwanted sub-system, e.g. the undesired response of a telephone line. The answer to the problem lies in the convolution theorem, which tells us that in the frequency domain this complicated operation of convolution is replaced by the simple operation of multiplication, and conversely the operation of deconvolution is replaced by division. In terms of poles and zeros, equalization means placing a pole over every zero and a zero over every pole, which pre-supposes that we are capable of synthetizing appropriate systems.

Digital methods give us two powerful ways of overcoming this important problem. Firstly, given pole-zero locations, synthesis is easier, we can often produce a digital filter which performs the equalization process. Alternatively if it is possible to treat the signal in terms of discrete blocks rather than continuously, we can perform the following process: take the Fourier transform of the output signal, divide it by the Fourier transform of the pulse response of the system (i.e. the frequency response), then take the inverse transform, and the result is the desired input signal. These methods can be of the greatest significance in the application of the more powerful microprocessors to measurements which of necessity have to be made through imperfect transducers, but the design engineer must always be aware of the important restrictions discussed above.

### 6.3.4.6 Matched filters

We have seen that the effect of a linear system on a signal can be found by taking the impulse response of the signal and the signal function, reversing one of them with a variable delay and finding the area under the curve represented by their product. This verbal interpretation of the convolution integral (§ 2.1.7) gives a clue to the idea of the matched filter, which is a means of detecting the presence of a pre-defined waveform within an uncorrelated (noise) signal. This capability is the solution to a small but important class of measurement problems. A matched filter for a given time-limited signal in the presence of white noise is a filter whose pulse response is the time-reversed version of the target signal. The nature of the convolution integral ensures that there is a positive peak in the output of the filter whenever the target waveform appears. This is a special case of the generalized matched filter which is obtained by the solution of the integral equation (Whalen 1971)

$$\int_0^T h_o(u) \, R_{nn}(\tau - u) \, du = s(T - \tau), \quad 0 \leq \tau \leq T \qquad (6.32)$$

where $h_o$ is the impulse response of the filter, $R_{nn}(\tau)$ is the autocorrelation (§ 2.1.8) of the background noise and $s(t)$ is the signal, of duration $T$, to be detected.

Only in the special case of white noise where $R_{nn}(\tau) = \delta(\tau)$, do we have the simple normalized result

$$h_o(\tau) = s(T - \tau). \qquad (6.33)$$

The matched filter is a prime example of a case where the use of digital filtering is greatly superior to the continuous variety. This is because the filter response is defined in the time domain and the synthesis of a continuous filter would require some frequency domain approximation to try to match it. In contrast the digital

## INTELLIGENT SENSOR CONCEPTS

matched filter is simply a non-recursive one with the coefficients set as the sampled values of the target function in reverse order.

An important case of a medical application of the matched filter is in detection of the *epileptic precursor* in EEG. This will be discussed below ( § 6.4.1).

### 6.3.4.7 *Waveform recognition and classification*

One of the more powerful techniques made available by the inclusion of intelligence within the the sensor is the recognition and classification of waveforms. This may be required for either positive or negative reasons, i.e. for the identification of a target waveform or the elimination of a given waveform known to pollute the incoming signal. The first necessity is for a transient capture scheme, i.e. a transient recorder implemented either in software (§ 6.2.2.2) or in hardware (for higher speed phenomena). The classification process involves the extraction of a number of characteristics (or metrics) from the waveform. If there are $N$ of these, then each waveform can be thought of as a unique point in $N$-dimensional space, and regions of that space are occupied by the different classes of waveform.

There is a variety of metrics that can be extracted from a waveform, e.g.
RMS value.
Number of zero crossings.
Dominant frequency.
Maximum slope.
Maximum slope of the slope.
Maximum slope of the envelope.
etc.

In the favourable case, where the regions corresponding to the various waveforms do not overlap, it is possible to make a positive identification of each occurrence of each waveform. As an example consider the case of the photomultiplier when it is used for detecting optical phenomena of very low intensity. There is a form of transient noise associated with photomultipliers that can be identified by certain characteristics. The intelligent photomultiplier system can eliminate such unwanted transients and concentrate on the target phenomena ( Buffam 1976).

### 6.3.4.8 *Doppler velocimetry*

Doppler velocimetry does not strictly measure velocity, but rather one component of it along the line joining the target object with the sensor.

Consider a wave of wavelength $\lambda$ radiating from a source and reflected back from a target object. If the object moves towards the source by a distance $x$ then the total path length of the radiation changes by $-2x$. This means that there is a change of phase by an angle $4\pi x/\lambda$. Thus if the original phase was $\omega t + \varphi$ the new phase will be $\omega t + \varphi + 4\pi x/\lambda$. If the object is moving with a component of velocity $v = dx/dt$ along the joining line the angular frequency of the returning waves, which is by definition the rate of change of phase, becomes $\omega + 4\pi v/\lambda$, and if the object is moving at an angle $\alpha$ to the joining line we have the characteristic Doppler equation

$$\omega' = \omega + 4\pi v \cos(\alpha)/\lambda \ . \tag{6.34}$$

It will be noted that we have not yet specified the type of radiation. It might be electromagnetic or ultrasonic. In either case the result stands, but we would not normally consider processing directly the incoming signal $A \sin \omega' t$, so we use the principle of the heterodyne by mixing it with the original frequency, $\omega$, to produce a difference frequency $\omega' - \omega = 4\pi v \cos(\alpha)/\lambda$. The basic problem with this scheme is that it does not preserve the sign of $v$, and if this is important we have to choose as a mixing frequency a frequency lower than $\omega$ by a sufficient amount.

In order to obtain velocity information directly form the incoming sine wave, we have to transform to the frequency domain. This can be done directly by DFT, (§ 6.3.4.2), but the signal may well be immersed in noise; in which case we can form the autocorrelation, which being an average eliminates the random component. Furthermore, by Fourier transformation we obtain the spectrum of frequencies (§ 2.1.8), which is directly representative of the spectrum of velocities. Thus the intelligent Doppler velocimeter not only enables the recording of single velocities but also can provide a complete spectrum where the velocities are distributed. This facility can be important in such applications as monitoring flow.

## 6.4 APPLICATIONS

### 6.4.0 Introduction

In this section we take a few examples of application areas which show how some of the fundamentals introduced earlier on are applied in real problems of sensing. Again the selection is not intended to be comprehensive, but is rather illustrative in nature. We are concerned here with principles rather than physical realizations, which we shall pursue in Chapter 8. Furthermore, in order to maximize the variety of these few examples, we shall start with some highly specific measurement requirements, where the fundamentals presented earlier are an essential element, and then go on to a couple of very broad discussions, in which the fundamentals are a constant background.

### 6.4.1 Petit mal detector

This important medical problem is useful as a an introduction to a general class of instrumentation problems in which the requirement is to detect the occurrence of a waveform of a particular type in the midst of a complicated signal (§ 6.3.4.6). Petit mal is a mild form of epilepsy occurring mainly in children, in which the overt

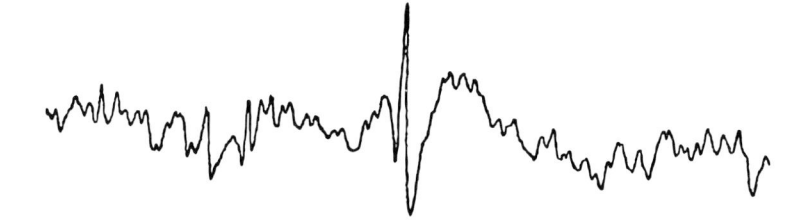

**Figure 6.15** The spike and wave complex – known to be an epileptic precursor.

symptoms are little more that a momentary loss of concentration. It is known to be associated with the occurrence of a particular waveform known as the spike and wave complex (figure 6.15) in the EEG (brainwave) pattern (Comley and Brignell 1981).

A first approach to the problem is to apply the simple matched filter concept, in which a filter whose impulse response is the time reversed form of the sought waveform is applied to the signal, so that a peak is reached whenever the signal occurs (§ 6.3.4.6). This simple approach can to some extent be successful, but it is based on the false premise that the background signal is white noise. In fact the EEG signal is not only non-white, it is also non-stationary (§ 2.1.8).

The solution to the problem was to use a dual processor system, in which one processor operated as a digital filter, while the other updated its stored coefficients according to the current state of the background signal. The process of calculating these coefficients is rather too involved to quote here, but it involved an elegant process of online deconvolution. It was originally implemented on a relatively large dual microprocessor system (Knight 1984), but the rapid changes in ASIC technology mean that it is now feasible to think in terms of a single chip solution. This would be a very useful contribution clinically, as it would allow ambulatory testing of children rather than having to attach them to large and intimidating equipment.

There are many other areas in which detection of particular waveforms is imperative, and there are generalized methods as described in § 6.3.4.7. The adaptive matched filter is potentially a powerful tool and a good example of why we seek to place more that one microprocessor on a chip. No doubt these multiprocessor techniques will be revisited in the light of the new technology.

### 6.4.2 Sensor array processing

Array processing has a general meaning in computer science, but here we are discussing the multiple signals that come from an array of sensors. Furthermore in this discussion we shall not cover the general area of processing signals from optical or ultrasonic arrays, as these are more in the nature of image processing, and while immensely important are somewhat outside the purview of the present text; for they involve complex techniques of parallel processing not yet applicable to the general sensor field.

Earlier ( § 3.2) we offered array processing as an alternative to the sensor-within-a-sensor to solve the cross-sensitivity problem. The classical area of application for arrays of sensors is the field of gas sensing. Here the problem of cross-sensitivity is so severe that it is spoken of as a problem of lack of *specificity*. Various materials change their electrical characteristics when exposed to certain chemicals, particularly in the form of gases and vapours, among them are oxides of zirconium, tin, zinc and many other metals, with the addition of various dopants. Also there are organic compounds, and particularly the phthalocyanines of certain metals, which exhibit similar properties. This is a very intricate area and largely outside the scope of the present text. Fortunately, however, there is an excellent multi-author text in the same series which deals comprehensively with the subject with copious references (Moseley and Tofield 1987) and another which deals with the mathematical nature of the processing (Moseley *et al* 1991). In the following those references are implied.

For our purpose, suffice it to say that by operating various lumps of material at various temperatures we can obtain different responses to different gases. There are, however, a number of profound problems.

First, none of these materials responds to only one gas, and the response to various gases is only marginally different, the difference also being a function of temperature.

Second, the response is not necessarily linear, and in some cases actually changes its sign as the concentration of the measurand gas increases.

Third, the base is not stable, in that the characteristics do not return to zero after exposure to gas, i.e. we have a severe hysteresis problem.

Thus in graphical terms we are trying to plot points with axes that are not orthogonal, scales that are non-linear and an origin that drifts about on the page. This summarizes the problems and illustrates why so much work has gone in to providing relatively few reliable operating devices.

In general the approach to this problem is to pretend that we have an orthogonal set of axes with linear scales and develop a more or less heuristic approach to distinguishing between various combinations of gases. This is not a task to be underestimated. Going back to our archetypal multivariable system( figure 2.1 ), when it was linear and orthogonal we could expect $N$ different inputs to produce $N$ distinguishable outputs, i.e. in the present terms $N$ sensors would allow us to evaluate concentrations of $N$ different gases. As soon as we have non-linearity and non-orthogonality this convenient arrangement breaks down. As with our non-linearity problem in analogue-to-digital conversion (§ 4.5.1) instead of having a one-one mapping we have a many-one mapping, which is to say **there is not necessarily a unique solution**.

There is an enormous variety of methods for tackling these problems, but there are a few common factors. First, we concentrate on changes of response rather that absolute values, which is analogous to the process of drift-tracking (§ 8.2.4).

Second, we scale the various inputs so that their variations are of similar orders of magnitude, i.e. one that shows a variation of parts per million is not swamped by one which varies by parts per thousand. Third, we treat the responses as though they were in fact orthogonal and of linear scale.

Thus we have an array of $N$ different materials (or similar materials at different temperatures). The response of each we imagine to be measured along $N$ orthogonal linear axes, so that each exposure to a gas or vapour mixture produces a point in the $N$-dimensional space spanned by the imaginary orthogonal basis. What we do next differentiates all the different techniques of pattern recognition as applied to this problem. Before exposure to the gas sample the system is at the origin of our space. After exposure it moves to some point away from the origin. Measurement errors will turn this point into a hyper-volume.

The operation of the animal nasal organ is not well understood, but it is likely that it is faced with similar daunting problems, and the question of odour sensing (the so-called electronic nose) is an important application. The techniques employed are various. Some are based on the computation of various statistics, such as cluster analysis, but there are many others ( the review by Gardner and Bartlett 1991 is recommended). In one the data are reduce to two dimensions by, in effect, projecting the points in $N$-dimensional space onto a single plane, and it is found that the points cluster in a manner that appears to be highly correlated with subjective descriptions of odours. Lack of space precludes further discussion here, but the references are recommended for readers who wish to delve further, and we shall revisit these matters in terms of physical realizations (§ 8.4).

### 6.4.3 Sensor response compensation: the load cell as an example

#### 6.4.3.0 *Introduction*

In this section we go into some detail in describing a particular compensation exercise carried out in the authors' laboratory shortly before completion of this text. It is of some value as it draws together a number of the threads from earlier in the book, and in particular highlights the importance of basic systems theory (§ 2.1). While we are concerned here with principles it is useful in this case to have an actual example, though we are anticipating the requirements of a subsequent section (§ 8.2).

As we have seen, the introduction of digital processing of progressively increasing power and decreasing cost has made substantial inroads into the problems of compensating for the natural defects of transducers. For example, the once intractable problem of non-linearity is amenable to solutions such as look up tables, while the robustness of digitally stored variables has permitted the tracking, and hence the compensation, of drift. One of the last bastions to fall has been the frequency compensation problem. This is partly due to the greater requirement for processing power, but also because of the difficulty of tailoring filters to individual devices. In the earliest days of computer-aided measurement (Brignell and Rhodes 1975,

Brignell and Young 1979) the available processing power meant that compensation for frequency response could not be carried out in real time and had to be performed by such methods as Fourier transforms (§ 6.3.4.5) but now small and inexpensive microprocessors offer the possibility of real time compensation by digital filters dedicated to a particular device.

The load cell is a prominent representative of a class of transducer that is particularly formidable in this respect. These are the devices in which the measurand contributes significantly to the parameters of the system, and hence to its dynamic response. Thus, as the load increases, both the natural frequency and the damping decrease, which leads to the paradoxical situation that, in order to determine the load quickly, we need to know the load before appropriate compensating filter parameters can be determined. We are, in fact, describing a non-linear system, since the relationship between an applied mass $m U(t)$ and the output voltage, say $y(t)$, is only additive when a static equilibrium voltage, say $y_s$, is established. The general aim in weighing systems is to reach $y_s$ in the shortest time possible. Traditionally the compensation method employed is to use mechanical damping; i.e. a single pole low-pass filter, to iron out the oscillatory response of the sensor system. Latterly this has been backed up by forms of digital smoothing in a microprocessor. Either way the time required for settling is several times the natural period of the sensor with its highest load applied. This 'sledgehammer to crack a nut' approach results in a weighing process that is unnecessarily slow, resulting in human irritation and loss of productivity.

This section is concerned with two contributions to the solution of these problems. The first is in the area of what might be broadly described as system identification techniques; though, in fact, we proceed directly to the identification of the compensating filter, with identification of the transducer system emerging as a by-product. The second concerns the use of adaptive filters to overcome the above paradox.

### 6.4.3.1 *The optimization approach*

The characteristics required for an intelligent sensor compensation technique point strongly towards an optimization approach (§ 6.2.2.4). They both require a set of numerical parameters which define the system, but the latter has two additional requirements - a measure of goodness (or cost function) for any particular numerical realization and a means of proceeding from one realization to a better one. A further relevant remark may be made with reference to figure 6.20. Consecutive transducers coming along a production line are not likely to differ a great deal in characteristics, so using the numerical parameters for the previous device mean that the initial guess is a good one, and the search process is thereby greatly truncated.

There are two main classes of optimization process – gradient descent methods and gradient free methods. In general gradient descent methods are to be avoided in transducer compensation applications, since noise in data acquisition tends to be a problem and the process of taking a gradient inherently amplifies high frequency noise (§ 4.5.4). A comprehensive review of available optimization techniques (Shi

# INTELLIGENT SENSOR CONCEPTS

1992) leads to a conclusion that the simplex method (Nelder and Mead 1965) is highly suitable for this application. Briefly the simplex in $n$-dimensional space is a figure of $n+1$ vertices, and the method prescribes a sequential process of reflection, expansion and contraction designed to approach the minimum of a cost function.

The choice of cost function is not always simple. First, it is conditioned by the choice of stimulus used in obtaining the system response. In principle any stimulus function containing all relevant frequencies is suitable, and the impulse is an obvious candidate. In the initial work for this investigation the impulse stimulus was used with some success (Brignell 1987a), with a cost function of the form

$$\frac{1}{y_{max}} \sum_{i=1}^{n} |y(i - i_{max})| + k |(y_{min})| \qquad (6.35)$$

where the largest sample $y_{max}$ occurs at discrete time $i_{max}$.

This is clearly zero for a perfect impulse, the final term being added to obviate negative going responses. There is, however, a lack of generality in impulse testing over the whole gamut of transducer applications. In practice this means applying the stimulus then immediately removing it. Usually the sudden application of a stimulus is not a problem, but the sudden removal can be. Consider, for example, the case where the stimulus is the presence of a chemical species. For these reasons of generality we decided to concentrate on the step function form of stimulus.

The choice of cost function for the step function test is not at first sight so clear cut. However, a thorough investigation by simulation (Shi and Brignell 1991) of four possible candidates established the superiority of what is perhaps the most obvious candidate – the error area as illustrated in figure 6.16. In fact a simple simulation showed little difference between the four different candidates until noise was added, when the divergence was dramatic (Shi and Brignell 1991). An added attraction of the error area as a cost function is that the presence of noise increases it, so the result of optimization is a nice compromise between ideal response and noise minimization.

There is also a choice to be made with regard to the array of parameters used to define the transducer, and hence the compensating filter. It is possible to use the coefficients of the polynomials in the $s$- or $z$-domains, but in this work we chose to use the co-ordinates of the poles and zeros in the $s$-domain.

In summary the optimization process comprises the generation of a step function in the relevant physical variable, which is applied to the sensor. The sensor output is acquired by means of a software transient recorder (§ 6.2.2.2) digitized and applied to a digital filter, initially with arbitrary coefficients, and the output of the filter is used to form the cost function by the subtraction of an ideal step and integration of the difference area. The simplex search is used adjust the roots of the digital filter to make the output more like the ideal step, and the process is repeated until further improvements are negligible. Finally, the digital filter is such as to produce the nearest obtainable approximation to an ideal step output, and furthermore the reciprocal of its transfer function is a model of the transducer system; so we have performed system identification as a by-product. The same principle can be applied to output transducers, or actuators, except that the order of connection is reversed.

### 6.4.3.2 Adaptive filtering

As we have observed, it frequently occurs that the physical variables that a sensor is designed to sense themselves affect the parameters of the sensor system. Indeed this can be used as a means of measurement, particularly with vibrating elements used in such applications as viscometry. Nevertheless, the phenomenon is on the whole disadvantageous. In much of the work on which this text is based, the load cell has been used as the typical sensor. In a way, however, this is an atypical sensor, in that the load itself is a dominant inertial parameter of the sensor system. Thus a filter realized by the above method is useless in practice as the sensor becomes a different system, with decreased damping and natural frequency, as soon as the load is increased.

Fortunately most load cells, like the majority of electromechanical transducers, conform to a simple second order model, and this gives a clue as to how the problem can be tackled. The system representing the load cell and its load can be described with some accuracy by the following second order differential equation (§ 2.2.6).

$$(m + m_o)\ddot{x} + \lambda \dot{x} + k x = F(t) . \quad (6.36)$$

Here $m_o$ is the effective mass of the load cell (somewhat less than the true mass, as it is constrained), $m$ is the applied load, $\lambda$ is a damping constant, $k$ is a spring constant and $x(t)$ is the deflection due to forcing function $F(t)$

In the complex frequency domain we have the response function $H(s)$, given by

$$H(s) = \frac{X(s)}{F(s)} = \frac{(m + m_o)^{-1}}{s^2 + \lambda(m+m_o)^{-1}s + k(m + m_o)^{-1}} . \quad (6.37)$$

The conjugate complex poles of this response are $a \pm jb$, where

$$a = -\frac{\lambda}{2(m + m_o)} \quad (6.38)$$

and

$$b = \left( \frac{k}{(m + m_o)} - \frac{\lambda^2}{4(m + m_o)^2} \right)^{1/2} . \quad (6.39)$$

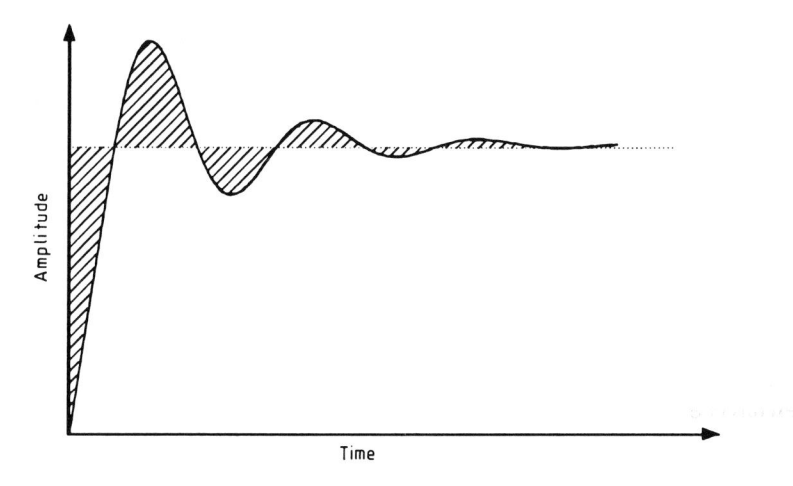

[Figure 6.16: Damped oscillation waveform with shaded error areas above and below the settling line]

**Figure 6.16** Illustration of the error area as a cost function for frequency response

A digital model for the system can be obtained by the matched z-transform in the general form

$$H(z) = \frac{C}{1 + Az^{-1} + Bz^{-2}} \qquad (6.40)$$

where $A = -2e^{-2aT}\cos(bT)$, $B = e^{-2aT}$, $C = 1 + A + B$ and $T$ is the sampling interval.

The clue to realizing an adaptive filter which will take account of the load is to observe that, as the load cell exhibits good linearity in static weighing terms, the final static output voltage, $y_s$, is directly proportional to the applied load. Thus we can recast the equations above in a form in which the parameters of the filter are conditioned by its own output level, $y$, just as the parameters of the sensor are conditioned by its input, $m$.

$y$ and $y_o$ now take the place of $m$ and $m_o$, while new empirical constants replace those above, including the proportionality between output voltage and deflection. Thus the compensating filter has zeros in the s domain given by $a \pm jb$, where

$$a = \frac{\alpha}{y} \quad \text{and} \quad b = \left( \frac{\beta}{y_o + y} - \frac{\gamma}{(y_o + y)^2} \right)^{1/2}. \qquad (6.41)$$

The roots in the z-domain are obtained by the above transformations and can be obtained equally in terms of empirical constants, which can be determined by a simple calibration process at two values of load.

The filter is now, of course, non-linear, but this is tolerable in view of our declared aim to reach $y_s$, which corresponds to $m + m_o$ in the shortest possible time, and as we have observed the weighing process itself is strictly non-linear. The theory of operation of the filter can be described verbally as follows. When the load, $m$, is initially applied the zeros of the filter correspond to the effective load cell mass, $m_o$. The poles due to $m + m_o$ dominate the overall system of filter and load cell, so an oscillatory response with overshoot begins. However, as soon as the output begins to rise, the zeros of the filter are constrained to follow a trajectory defined by the equations above ( see figure 6.17). Eventually the output reaches a level equal to the correct static output, $y_s$, and the filter is exactly correct for the applied load, $m$ ( i.e. its zeros annihilate the poles of the load cell plus load). Thereafter the output voltage and the filter coefficients cease to change until there is another change of load. It follows that the trajectory followed by the output voltage lies somewhere between the raw oscillatory response, due to $m + m_o$, and the ideal step, until the plateau is reached. Thus ideally the correct final response can be reached well before the first oscillatory quarter-cycle of the raw response is completed. Since the model is bound to be imperfect to some extent a simulation study was carried out prior to experimental verification of the principle.

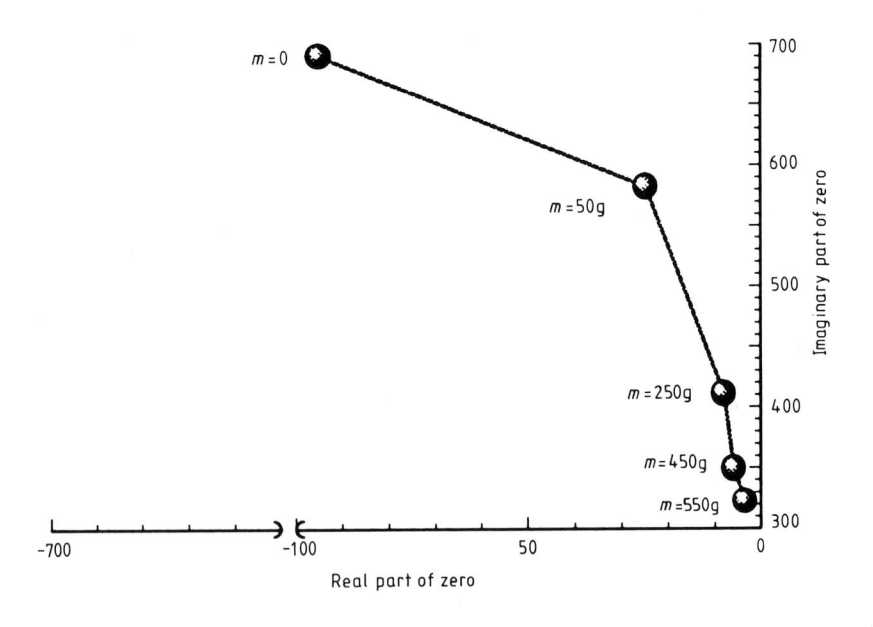

**Figure 6.17** The trajectory of one of the complex poles of a load cell as the load increases. Note that the two axes have different scales.

#### 6.4.3.3 *Simulation of the adaptive filter*
Equation (6.40) provides a digital model for the load cell, and measured values from the optimization process applied to a real device allow its behaviour to be simulated for any load. A simulated step change in load modifies both the input to the device model and its poles. When the step is applied the filter zeros are initially set at the point on the trajectory corresponding to the previous static load, and as the oscillatory response to the load change begins they move along the trajectory of figure 6.17 in a direction dictated by the sign of the load change and hence the filter output voltage change. If the filter operates in the way surmised in § 6.4.3.2 then the filter zeros should settle over the load cell poles in less than a quarter of a cycle of the oscillatory response.

In fact, simple simulations confirmed the principle of operation. In order to make the assessment more realistic, however, it is necessary to add noise, since the action of the filter in flattening the spectrum of the overall system has the effect of re-emphasizing system noise. Figure 6.18 represents one of many simulations. Here, in simulation, a mass is applied and the load cell goes into its oscillatory response. The effect of the adaptive filter is remarkably evinced. As the output of the filter rises it adjusts its poles to get closer and closer to those corresponding to the applied mass, which means that it rises faster than the filter input, and is closer to the ideal step input to the load cell. It is also clear that noise is a residual problem, which will have to be dealt with in a way that causes minimum deterioration in the speed of response. Further simulations showed that the adaptation was even more efficacious with a downward step in mass,as it is dominated by the leading zeros of the filter. This differential treatment of rising and falling edges is an illustration of its non-linear nature.

#### 6.3.3.4 *An adaptive filter for noise suppression*
The frequency response compensation process leaves a residual noise problem. The obvious solution to this is to use a single pole low-pass filter of the form

$$z_i = (1-b) y_i + b z_{i-1} \tag{6.42}$$

where the time constant is given by

$$\tau = -\frac{T}{\lambda} \log_e (b) . \tag{6.43}$$

Figure 6.18 (b) shows the effect of applying such a filter with a time constant of 0.2 seconds. The noise problem is solved, but at the sacrifice of some of the speed gain. We can improve the situation by taking liberties with linearity as before, so that the time constant is made dependent on how close the system is to equilibrium. The following scheme was developed for this purpose.

First the absolute value of the difference between the input and output of the smoothing filter is formed:

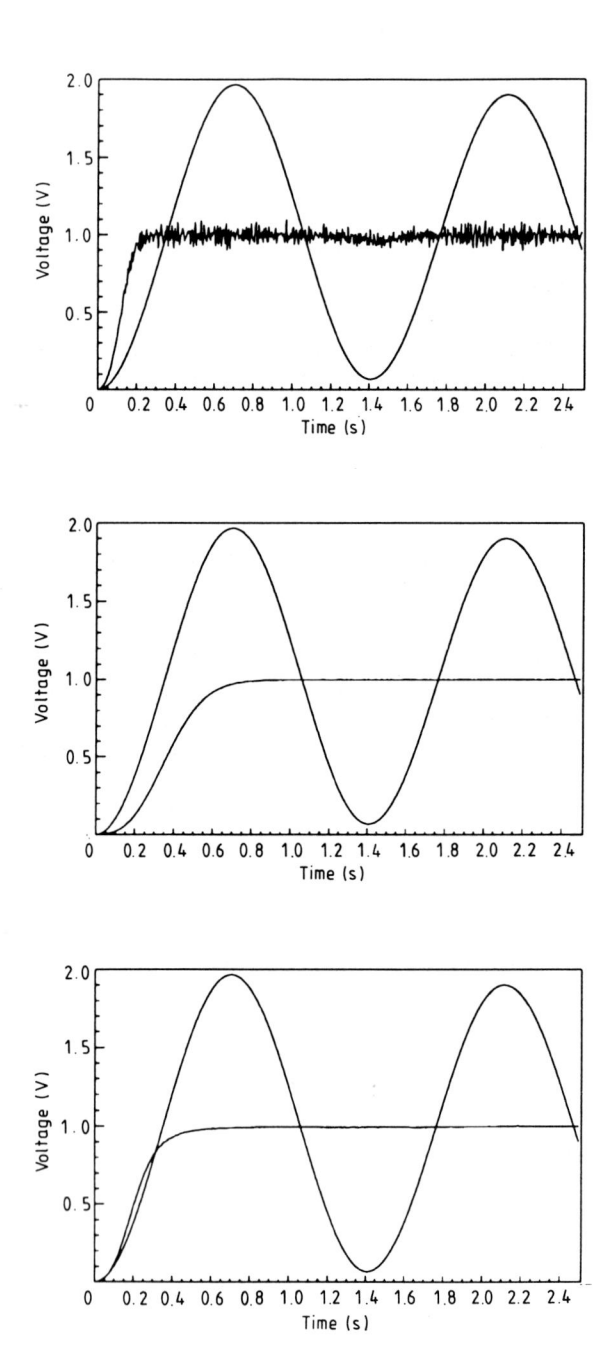

**Figure 6.18** Simulation of load cell adaptive filtering (a) effect of the adaptive filter (b) with noise filter of fixed time constant (c) with adaptive noise filter.

$$\Delta = |z_i - y_i| \tag{6.44}$$

and the time constant is obtained from

$$\tau = \frac{1}{100\Delta + 5} . \tag{6.45}$$

In this way the time constant is long (0.2 sec) as equilibrium is approached, but short when the equilibrium is disturbed. This achieves the objective of a rapid transition to the plateau but adequate smoothing of the plateau once it is reached. Figure 6.18 (c) shows the effect of adding such an adaptive noise filter to the output of the adaptive compensation filter. Of course, the numbers in equation (6.45) have to be adjusted to circumstances. This technique is an analogue of a well known one in continuous electronics, in which the resistor in an *RC* integrator filter is replaced by two parallel diodes connected inversely.

### 6.4.3.5 *Experimental results*

In order to test the techniques developed in a real weighing situation, two load cells were selected. The first was a standard 1010 type, but with the mechanical damping material removed. The second was a tri-beam load cell developed in the authors' laboratory (§ 8.2.2). The latter posed a particularly demanding problem of compensation, as it is a planar thick-film device with little mass and natural damping. One problem of experimental technique is worth noting. It is difficult to add weight to a system without bounce; so most of the tests were made by first applying the load plus a small test weight, then suddenly removing the test weight by a sharp downward blow on a simple lever to which it was attached. Thus the excitation was, in fact, negative going, and the signal was inverted for the purposes of optimization and plotting, but the dynamic situation was determined by the chosen load.

Fortunately in practice it is found that a simplified model of the pole behaviour is sufficiently accurate, and the poles for the 1010 load cell follow a trajectory well defined by:

$$-\frac{7.018}{y_s + 0.07392} \pm j \frac{515.1}{\sqrt{y_s + 0.5569}} \tag{6.46}$$

where $y_s = 0.0036m$, which is, in fact, the locus plotted in figure 6.17.

The locus of equation (6.46) was used to control the zeros of an adaptive filter as outlined in § 6.4.3.2, though, of course with the output of the filter substituted for $y_s$. Figure 6.19 (a) shows a typical result for the output of the adaptive filtering system when a load of 50 g is applied to the 1010 load cell, while figure 6.19 (b) shows the same cell with a load of 450 g. In each case the amplified raw output of the load cell is superimposed on the same axes. It will be observed that, despite the great differences in frequency and damping for the two cases, the output waveform

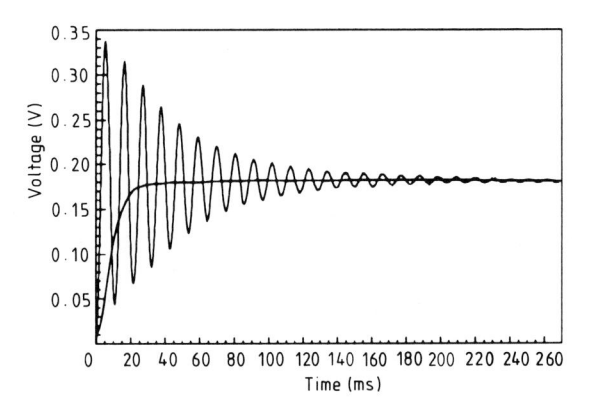

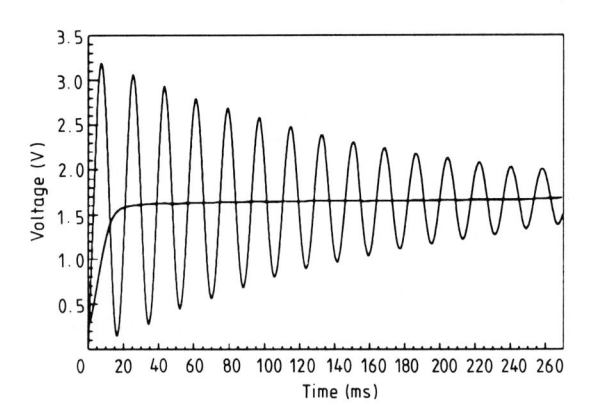

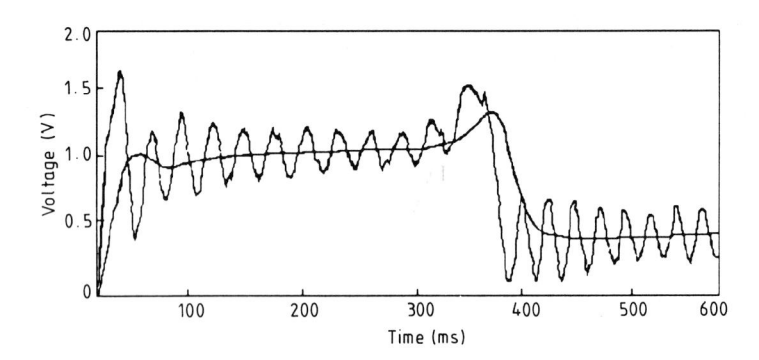

**Figure 6.19** Experimental results with adaptive filtering of the load cell response (a) the 1010 load cell with 50 g applied and (b) 450 g applied (c) response of the tri-beam load cell with 500 g applied then 200 g removed.

is virtually identical, and the equilibrium value of load is established in a few tens of milliseconds. This is a speed of weighing which is more than adequate for human observation and for many automatic systems. Similar results were obtained over the entire span of the load cell.

The tri-beam load cell is rather more resonant and noisy than the more conventional type, but as it has the advantages of being very cheap and robust, it would be very useful if it could be fully compensated. Figure 6.19 (c) shows one result which sums up many tests performed on the device. In this a 500 g weight was added and then 200g removed a fraction of a second later. Although the time to achieve equilibrium is nearly an order of magnitude greater than in the former case, there is still a remarkable improvement on what was virtually an intractable signal.

### 6.4.3.6 *Discussion*

What we have attempted in the above is an inversion of a non-linear process. The results are imperfect but still a substantial improvement on what is almost an unusable waveform, which is corrected by traditional methods only for a significant sacrifice of speed of measurement. Although weight sensors are somewhat untypical in the degree to which the measurand is a parameter of the system it is by no means an uncommon phenomenon. The stiffness of a pressure diaphragm, for example, is a function of the applied pressure.

The other factor in this work that might be considered somewhat special is the aim to measure the establishment of a new plateau as quickly as possible. Again this is by no means a unique requirement. Furthermore, the method does not rule out the tracking of continuously changing weights, such as in the filling of a hopper, but the accuracy of the tracking will suffer if the rate of change is such that it is measurable within the settling time of the system (some 30 milliseconds for the 1010 load cell). Nevertheless, this is a considerable improvement over what can be achieved by the traditional methods of gross electrical and mechanical linear low-pass filtering.

These considerations point to an important factor in intelligent sensor realizations. It is not only the nature of the sensor that matters, but also the nature of the measurement process in which it is involved. One of the factors that are unique in weighing is the phenomenon of 'bounce'. Indeed, it is quite difficult to apply a sudden load change without bounce occurring. Simulation studies showed that the adaptive filters quite accurately tracked a simulated bounce waveform. Nevertheless, it remains a restriction that the weight cannot be accurately determined until bouncing has ceased.

It is always an open question as to how far to go with the compensation process. No doubt the results here could be improved by further trimming of the processes, but it is doubtful whether the extra processing time can be justified. It is, for example, undoubtedly true that the second order model is imperfect, and when the dominant poles and zeros neutralize each other recessive higher order responses are expressed. In the above procedure, however, they are suppressed by the adaptive noise filter.

# 190 INTELLIGENT SENSOR SYSTEMS

Tests show that it is possible to characterize an individual load cell by applying the identification process at just two values of load, since the behaviour adheres quite well to the simple second order model developed in section § 6.4.3.2. This confirms the feasibility of manufacturing intelligent load cells which can be quickly calibrated on the production line as envisaged in figure 6.20.

## 6.4.4 The intelligent building

The design of buildings has been an area that is crying out for the application of intelligence, in more senses than one. Much of this text was prepared in offices in a completely new building that are uninhabitably hot in the height of summer, yet in the winter the temperature has to be controlled by opening the windows, as the heat sensor is on the cool side of the building. Kilowatts of heat pour out from each office, whether it is occupied or not, enough value to pay for a proper control system in no time. Pilfering is carried out, probably by strangers who wander in off the street out of normal hours. Fire regulations require the closure of internal doors, even when there is no fire, thereby undermining the group interaction that is vital to a modern academic department. Why is this?

The design decisions were made by people who are not capable of making the right technological choices. Architects, even after a long training, are unable to calculate the stress in a simple beam, let alone tackle the thermal design of their creations. They regard themselves as artists, to whom the comfort and financial health of their victims are secondary issues. The architects talk to buildings officers who are equally uneducated in the vital technological areas. Finance is under the control of politicians, who will happily squander vast sums of money in future running costs to make the initial capital expenditure look more modest.

### 6.4.4.1 *Environmental control*
The most important economic consideration in building control is that of temperature (§ 3.3.2). This is the potential variable that human's perceive and which affects their ability to perform their everyday functions, but as we have seen, as in all physical systems it is linked to a flux variable, in this case heat. Moving heat around costs a great deal of money, whether it is a net removal, as in tropical climes, or a net addition as elsewhere. The ratio of the two variables is an impedance. This may be counted as linear for small temperature differences (cf. Newton's law of cooling) and it comprises a resistance and a very large capacitance - in a crude model one of each, while in a more refined models complex network. Thus, to a first approximation, the basic system is effectively a simple single pole one

$$H(s) = \frac{a}{s+a} .$$
(6.47)

The impossibility of adopting our technique of response compensation (§ 6.3.4.5) in such a situation is revealed by considering the inverse response to a step

$$\frac{1}{s}H^{-1}(s) = \frac{a}{s} + 1 \tag{6.48}$$

which is an impulse plus a step (equation (2.10)), meaning that we have to supply an infinite amount of energy in an infinitesimal time to change the temperature and then a constant amount to maintain it. The compromise reached to overcome this impossible requirement characterizes the particular solution. In Victorian times the housemaid got up before the rest of the household to stoke up the fires with coal. In more modern times the central heating timer switched on a simple on-off controller, the sensor often being a relatively crude bi-metallic strip.

The intelligent approach to temperature control involves a careful balance between the economic and human considerations. Heat is a low grade energy, in that all other grades of energy tend to degrade towards it. Electricity, on the other hand, is a high grade form of energy, in that you can use it to do almost anything. It is therefore irrational to use electricity on a large scale directly for heating, though local political conditions may distort markets in its favour. In order to achieve economic control of temperature it is necessary to combine a number of considerations:

(a) the likely demand
(b) the actual demand
(c) the external temperature.

The likely demand can be predetermined and preprogrammed. It is conditioned by the time of the year, the day of the week (taking into account national holidays) and the hour of the day.

The actual demand is more subtle. It is, of course, largely determined by the external temperature, which can differ substantially on two sides of a building, a point which passes by many architects, but there are other subtle considerations. Is it sensible to keep rooms at temperature if they are unoccupied all day? If we let them cool down, how does it affect the comfort in adjacent rooms?

We have stated the fact that the external temperature is a factor in the actual demand, but equally important, and largely ignored in simple control systems, is its rate of change. It does not take a great deal of processing to predict the time at which demand will cease as the external temperature rises in the morning, but simple systems fail to do this and the resultant oversupply causes the occupants to take matters in their own hands by throwing open the windows. The resultant energy wastage is a great expense, and alone would justify the cost of a proper control system.

Temperature is not, however, the only environmental factor in human comfort. Humidity ( § 3.3.3.5) is also very important, particularly in the tropics, and we have seen how it can be measured, for example, by measurement of the change of resistance of a hygroscopic material. Again measurement of the external humidity is important. Air drying and cooling are expensive processes, so minimizing them makes economic sense.

Air quality is the third major factor in environmental control. Static air soon becomes stale and unpleasant with human occupation. 'Unpleasant' is a subjective term, and its use reflects the fact that evolution has given us an instinct to avoid such conditions and a sense organ to warn of their existence. Recycling air is an obvious economic issue in heating and cooling, and the objective of control is to reconcile economy with comfort by judicious mixing of new and recycled air. The simplest measure of air purity is the concentration of the major product of human respiration, carbon dioxide, which is most simply determined by infra-red absorption, while the search for reliable solid state devices continues.

There are other air contaminants. Dust is one which can be controlled by the circulation of filtered air. In special circumstances where it is likely to be a problem a working sensor can be based on the scattering of indirect light by the particles. There may also be other special circumstances where particular processes are carried out which produce known contaminants, sometimes poisonous or carcinogenic or even explosive. In such cases special chemical sensors will be required, and we note the problems of specificity with such devices (§ 6.4.2, § 3.3.3).

### 6.4.4.2 *Fire*

Fire detection self-evidently is a vital aspect of building control. The most obvious way of detecting fire is by way of temperature, but measuring that on its own is insufficient. What is important is the difference in temperature between one site and a nearby one. This is one of the areas where sensor networks come into their own (§ 7.1). The alternative, or more wisely adjunct, method is smoke detection. Smoke sensors are based on either optical or ionization techniques. Optically they rely on dispersion or absorption of light by the smoke particles. Ion detectors rely on the ion current produced by an alpha-ray source, which is reduced by the presence of smoke particles as the ions become attached to them and the current is reduced by the reduction in mobility.

The advantage of an having an intelligent sensor system in fire detection and control is substantial. The sensors can be self-checking. For example, the thermal fire detector is one of the cases where the self-check routine (§ 6.2.1.3) can be extended to the primary sensing element itself, by incorporation of a heater in the housing of the device, and similarly an extra light source enables the optical form of primary sensor to be self tested.

Digital communication is also a valuable addition. Early warning can be given of a potential fire even at a stage where there is insufficient information for a positive decision. Once the fire has started it can be tracked, and if a loop topology is adopted (§ 7.1) even the progressive destruction of the network itself can give information on the progress of the conflagration.

### 6..4.4.3 *Security*

Intruder detection is the main reason for the development of people-detecting sensors, but not the only one. It is wasteful to apply the whole range of services like those mentioned above to a room that has been empty for some time. Even switching off the lights in unoccupied rooms can produce significant savings.

There is a wide variety of methods for sensing the presence of human beings. The simplest are pressure pads, which issue a signal when trodden on. This idea may be extended to cables where the insulant is a piezoelectric polymer. Laid out in huge loops round buildings, they not only indicate the presence of an intruder but, by means of signal transmission time differences, can indicate a location as well. This is not a trivial task of signal processing, as the signals are irregular in form.

As warm bodies, human beings radiate in the infra-red region, a phenomenon exploited in inexpensive external intruder alarms. Infra-red can also be used to provide invisible beams, so that signals are issued when the path to a detector is interrupted.

A very powerful form of people sensor is provided by setting up a pattern of ultrasound standing waves which are disturbed by motion within the region covered.

Electronic keys enable the system to determine just who is present in each area. These may range from simple punched cards as favoured by many hotels, to magnetic stripe cards and even infra-red devices which provided two way handshake checks with changing codes for the ultimate in levels of security.

The important contribution of intelligent sensor systems to security is provided by the network, which we shall discuss in the next chapter. The information made available to the centre can be exploited in a number of ways. For example the system can learn the behaviour pattern of the building's occupants and warn security staff when some unusual activity is taking place. If a villain tries to buck the system by cutting cables, the system can identify the position of the cut within two sensor locations (§ 7.1).

It is clear that all these functions – environmental control, fire control and security control – can be exercised through one two-wire multidrop bus system (though for extra security we may provide for duplication, § 7.1). As always, the bus itself is a precious resource, and a strong motivation for using intelligent sensors is the data condensation that can take place before transmission, as well as the continual performance of self-check routines.

*6.4.5 Automotive applications*
Consider the fuel indicator in the traditional automobile. It is a byword for crudity of instrumentation, and a tribute to the tolerance of motorists. The coarseness does not derive from the simplicity of the sensor (traditionally a float operating a potential divider) but is a classical case of not measuring what we want to ( the volume of fuel) but measuring what we think we can (the depth of fuel). Fuel tanks are of odd shapes as they have to fit into a space determined by other design compromises. The intelligent sensor answer to this problem derives from the simple observation that, however complicated the shape of the vessel, **the volume must be a monotonic function of the depth**. This immediately leads us to the Look Up Table as a solution (§ 6.2.2.1). This is an excellent example of the simplest of signal processing techniques turning an unsatisfactory sensor into a satisfactory one.

The detailed practicalities of intelligent automobile systems are dealt with in detail elsewhere (Ohba 1992), which frees us to consider the important principles involved. The automobile can be divided into a number of sub-systems:

1) Supply (air and fuel)
2) Driver control input
3) Ignition and combustion
4) Power train
5) Exhaust
6) Suspension
7) Passenger chamber environment
8) Internal instrumentation
9) External indication

#### 6.4.5.1 *Supply*
In general the objective of air-fuel flow control is to maintain the conditions at which total combustion takes place without heat being wasted in excess oxygen. These ideal (stochiometric) conditions occur when the air-fuel ratio is slightly under 15:1, but they would normally be maintained with slight excess of air to obviate the production of carbon monoxide through partial combustion. There is a variety of ways of measuring the mass flow of fluids, as we have seen in § 3.3.1.4.

#### 6.4.5.2 *Driver control*
The driver makes demands on the other sub-systems by means of the manual controls at his disposal. Depression of the accelerator can be measured by means of a displacement sensor, often a potential divider of, say, conductive plastic with a metal brush wiper, though non-contact methods would give better life and reliability (§ 3.3.1.2). The accelerator creates a demand for air-fuel and if one is controlled the flow of the other follows.

Steering requires a rotational displacement sensor, again preferably of the non-contact type, such as an optical encoder. This information is used to control the wheels (either two or four) and the response allowed is varied with the ground speed of the vehicle (Ohba 1992).

Automatic breaking systems (ABS) can be of greater or lesser complexity. The main problem they have to deal with is optimizing the stopping time by taking into account wheel slip. The human operator, in an emergency, is likely to over-demand on breaking, which introduces the danger of skidding. In order to achieve optimal control it is necessary to have separate measures of wheel speed and ground speed, and when these do not match slipping is occurring. The former can be measured by a non-contact rotation sensor, either optical or magnetic, while the latter is perhaps best measured by doppler velocimetry (§ 6.3.4.8). The wheels can be separately monitored so that during maximum breaking each is kept on the edge of slipping.

Other controls are of a simpler nature and include indicator lights, windscreen washers etc.

### 6.4.5.3 *Ignition and combustion*

The firing cycle of an internal combustion engine is a complicated control problem, which was dealt with traditionally by a relatively crude and approximate mechanical system. The optimum condition to be maintained is just below the point at which knocking (spontaneous ignition) occurs. A knocking sensor is basically an accelerometer (§ 3.3.1.3). The ignition timing is related to crank angle, and its optimum settings depend on a number of things, mainly engine temperature and speed. This again is where the Look Up Table (§ 6.2.2.1) comes into its own, and the entries in the table can be evaluated with some precision by reference to engine test-bed data. This account is, of course, over-simplified, but it will suffice for present purposes.

### 6.4.5.4 *Power train*

The need to control the power train derives from the fact that the internal combustion engine has a torque speed curve that is relatively constant. This means that power ( the product of speed and torque ) is low at low speeds; hence the need to change the gear ratio. Good control of gear shifting affects the comfort of the ride, the fuel efficiency and the wear and tear. Torque in rotating shafts is one of the more difficult sensing problems, but to ensure optimum control it is a necessary one. Various attempts have involved strain gauges with magnetic coupling and optical monitoring of the strain distortion of the shaft itself. Rotational speed is, of course, a simpler matter.

### 6.4.5.5 *Exhaust*

There are two reasons why there is a need to apply gas sensing to the exhaust. First, the presence of oxygen indicates the efficiency of combustion, so there is a requirement to monitor this gas and maintain it at a desired low level. Secondly, there is the question of pollution. Because this is a serious social and health matter there is growing legislation controlling emissions of the main pollutants (oxides of nitrogen and carbon monoxide). It is therefore necessary to monitor these gases to ensure that there has not been a failure of air-fuel control or of catalytic converters. Again this boils down to gas sensing with all its attendant problems (§ 3.3.3, § 6.4.2). It is also desirable to monitor the temperature of exhaust gases, as catalytic converters can be degraded by overheating.

### 6.4.5.6 *Suspension*

The suspension is a major subjective factor in motoring. We can think of the suspension as being a typical second order mechanical system (§ 2.2.6, § 3.3.1.3) and we have two parameters at our disposal, the spring and damping constants. Like our load cell (§ 6.4.3.2) the suspension suffers from the problem that the vehicle load contributes to the inertial parameter of the system, so that as the load increases the natural poles move along a trajectory, like that of figure 6.17, reducing the damping and natural frequency of vibration. Without going into detail we can say that the optimum setting of the suspension depends on the vehicle speed, the nature of the road surface, whether the vehicle is accelerating, retarding or cornering. Also the suspension control can be used to set the vehicle height, which should be high for low speeds over poor surfaces and low for high speeds over good surfaces.

Full suspension control requires accelerometers, and sensors for force and displacement. It should be remembered that the tyres are part of the suspension system, and tyre pressure is an important safety factor. Measuring individual tyre pressures produces the same communication problem we have with measuring the torque on rotating shafts. That of powering the devices and getting back the information. A strong possibility for monitoring the health of the suspension system including the tyres is to treat the road surface irregularities as an input signal to the system. The front and rear suspensions should receive almost exactly the same inputs with a delay determined by the speed. Cross-correlating them produces a peak which is a measure of road speed, but if their spectra are different there is an indication of something wrong. This is an avenue for useful signal processing research.

#### 6.4.5.7 *Passenger chamber environment*
Passenger comfort presents all the problems of building control with the addition of problems caused by the motion. Temperature control is relatively simple. The suspension can affect comfort and travel sickness, which are subjective and therefore require some form of human selection of the nature of the ride. Noise and vibration are irritations which could once only be dealt with mechanically, but now are subject to control by electronic cancellation methods.

#### 6.4.5.8 *Internal instrumentation*
Electronic display methods are now so versatile that panel displays in vehicles are wholly determined by the human requirements. The ergonomics of such displays are, on the whole, beyond the scope of this text. In general digital displays are not found to convey relative information as efficiently as analogue ones, so we end up with a compromise in which discrete display elements are combined to imitate continuous ones. Of great importance is the indication of fault and emergency conditions, which can be emphasized by the use of colour and flashing.

#### 6.4.5.9 *External indication*
External indicators are important safety factors, particularly for turning and reversing. Self-test is a requirement, and a simple form would be a check on whether bulbs are actually taking current, though a safer test would be via a photodiode to ensure that they actually are illuminated or off at the appropriate times.

#### 6.4.5.10 *The whole system*
We have described here, very superficially indeed, a complicated system. Some parts of it operate on relatively slow signals, while others operate at high speeds requiring local control loops. One of the features of traditional vehicles is the elaborate cable harness, which is effectively a star network with separate wires

going to each sensor or actuator. It is a major cost component and the plugs and sockets are one of the most common causes of awkward faults.

The intelligent system approach is to have just one power bus and one communications bus going round the system (we can combine these, but it is debatable whether the cost savings justify the design contortions and possible extra failure modes). If the bus is addressed at both ends we can add a measure of fault location and tolerance (§ 7.1).

Consider the simple case where the driver indicates to turn left. The communications processor sends out messages to the front, side and rear left indicator lights to turn on. They recognize their addresses, turn on, and then refer to their inbuilt photodiodes to ensure that they are illuminated before sending back messages to say that they have obeyed the command. If any of them has failed, a warning message is returned which initiates visible and audible warnings to the driver. This is all done on a single communications bus that goes right round the vehicle, and takes a fraction of a second, the delay being therefore humanly unobservable. On and off messages are send at a strictly controlled frequency until the steering sensor indicates that the steering wheel has returned to centre.

If there is a minor collision which causes the bus to be broken at some point, the central computer sends out test signals to all devices from either direction, indicates to the driver that there is a fault and where it is, then reconfigures itself so that each device is only addressed from the appropriate direction. Such test signals are sent out continually whether the device is in use or not. Unless the damage is of a major kind the vehicle is able to continue in use until a repair is affected. The system is able to advise the repairman of the nature of the repair required.

The automotive market is a special one with unique constraints on cost and performance. In many ways, however, it does exemplify the requirements which lead to the technology we are concerned with in this account. The above brief verbal revue shows that it is an active test bed for these new ideas.

## 6.5 PRODUCTION

### 6.5.0 Introduction

The wide ranging subject matter in this chapter on concepts emphasizes the multi-disciplinary nature of the subject of intelligent sensors. All these fancy ideas, however, are nothing if they do not lead to competitive products. The two most significant factors in competitiveness are performance and price. There is plenty of evidence in these pages to support a contention that the intelligent sensor approach can greatly enhance performance. Indeed, it facilitates forms of measurement that are quite impossible by traditional methods. It is necessary, however, necessary to pay some attention to the question of production costs. This was the very Achilles' heel attacked by critics of the whole idea when it was first proposed. What, they observed, could be more fatuous than putting an expensive microprocessor in a cheap sensor? They were, of course, guilty of looking at the state of the art rather than economic trends. This question of trends is an important one, to which we shall return in Chapter 8.

### 6.5.1 Production costs

The well known division of production costs into fixed costs and variable costs is important in all manufacturing industry, and is particularly significant for sensors. Although there are a few application areas where large production runs occur (e.g. the automotive industry) on the whole small or medium runs are more the norm. We have discussed in Chapter 5 the various enabling technologies which facilitate the production of intelligent sensors. At the time of writing this text the dominant production technology had moved away from printed circuits (§ 5.1) to thick-film (§ 5.4.1), but, as we shall argue in Chapter 8, the trends are still continuing and appear to be moving more towards direct realizations on silicon. A major element in fixed costs is design, which of necessity relies on highly skilled man-power, and the mitigation of this factor is also something we shall address in the later chapter.

If, however, one pays a visit to a traditional sensor manufactory, one is struck by the dominance of final trimming of the sensors in the overall activity, and hence the costs. This therefore is the aspect that we shall give special consideration in this final discussion of intelligent sensor concepts.

### 6.5.2 Trimming

In § 3.2 we identified four different classes of compensation. The second of them, tailored compensation, corresponds to the classical methods of trimming employed in sensor manufacture. The third class, monitored compensation, is more a product of the introduction of intelligence into sensors, but it also involves an element of the trimming process, since the models operative for particular sensors will be individualized. If, for example, we employ the symmetry of the Wheatstone bridge (§ 3.2, § 4.1.1) as our primary form of structural compensation, the temperature compensation process will be prescribed by a residual and arbitrary error function, however systematic might be the temperature characteristics of the uncompensated sensor elements.

In various parts of this text we have used the precision load cell as our archetypal primary sensor, as it reveals many of the facets of the compensation problem that we need to address. A visit to a load cell manufactory in the 1970s would reveal many of the problems of the traditional approach. The actual manufacturing process comprises the machining of load cell billets and the affixing of strain gauges, simple and relatively cheap processes. Subsequent to this, however, comes a relatively man-power hungry process of production trimming, in which tests are made for

## INTELLIGENT SENSOR CONCEPTS

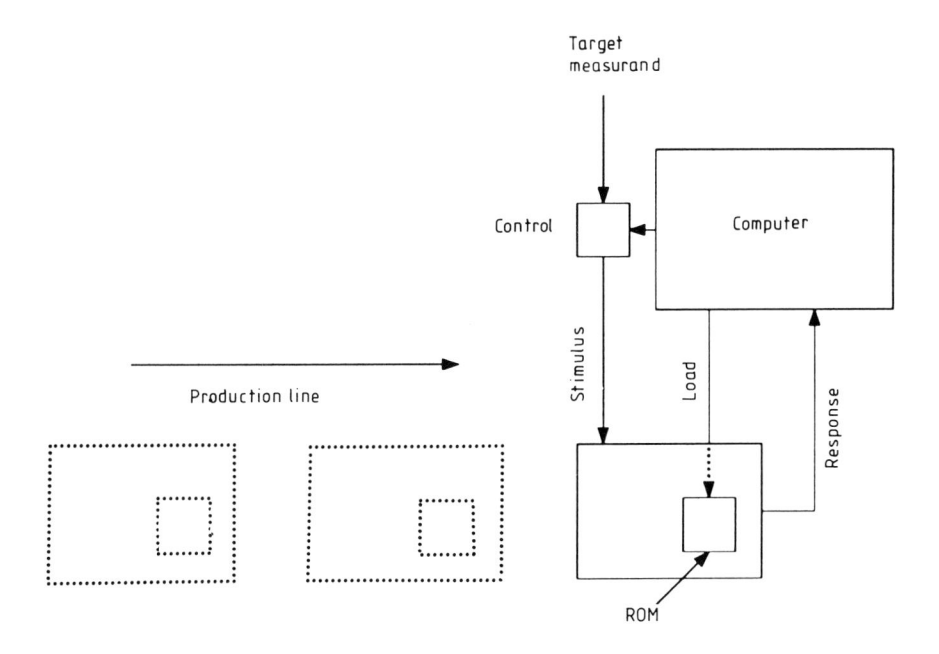

**Figure 6.20** Diagrammatic representation of intelligent sensor coming consecutively along a production line and being trimmed by a process of downloading coefficients.

temperature and eccentricity errors to permit subsequent human adjustment by means of addition of resistance wire and removal of metal by filing. These processes account for a major part of the cost of the sensors. In the intelligent load cell ( § 8.2) the approach is somewhat different.

### 6.5.3 Intelligent sensor trimming

By this stage in this treatment the reader may have noticed a common thread in the methods of compensation offered as a component of the intelligent sensor approach. This is that, however complex the process, it is individualized by means of an array of numbers. This is no accident. These numbers might be the coefficients of a digital filter (§ 6.3.1) or polynomial, the entries in a look up table (§ 6.2.2) or a simple stored offset. Whatever they may be, they are numbers that can be down-loaded into a ROM as the result of an automated test in the final stages of production.

Figure 6.20 is one of the most important illustrations in this book, since it demonstrates this process of specifying corrections by the down-loading of coefficients. There are, of course, a number of extra requirements. Interfering parameters (and especially temperature) have to be controlled. The target variable also has to be controlled, and this can be one of the most difficult problems, especially in frequency compensation (§ 6.4.3) where ideal waveforms need to be generated.

The concept of a sequence of sensors coming off the production line is an important one; for, though there may be a great deal of variation over the whole production range, it is unlikely that there will be a great variation between adjacent devices. Therefore, in the use of indirect software structures (§ 6.2.2.4) such as optimization (§ 6.4.3.1) the coefficients for the previous device are likely to provide a good first guess for the present one, thereby truncating the process and speeding up the production line. A further major advantage of the sensors being intelligent is that it is not necessary to provide accurate measurement of the environmental parameters, as the sensor can deal with these in its own internal units (§ 8.1.5).

It is, of course, necessary not to get carried away with the purity of the concept. As an example of a compromise approach it is quite feasible for a primary sensor manufacturer to provide a sensor with an accompanying ROM which defines its compensation coefficients, so that the instrument manufacturer may incorporate this in his own equipment with its own processor.

There are many other refinements of these ideas, which space precludes, but the above discussion will point the reader in the general direction.

### 6.6 STIMULUS-RESPONSE SENSORS (UNORTHODOX STIMULI)

Before leaving this central theme of intelligent sensor concepts it is worthwhile to mention a form of sensing that has hardly been exploited. This is stimulus-response sensing, with special reference to the capability of digital systems to generate forms of stimulus that are impossible to achieve by classical means.

Stimulus-response sensors obtain information about the environment or a target system by applying a physical stimulus and examining the response. Familiar examples are radar and non-destructive testing by ultrasonics. Normally the stimulus takes the form of one of the standard functions (impulse, step or sine), and as we have seen in Chapter 2, for a linear system each response is derivable from the other. Intelligent systems, however, offer a completely new approach which has rich possibilities, by allowing unorthodox stimulus functions to be generated. Let us briefly examine just two examples which illustrate the point.

Figure 6.21 (a) illustrates one such function, the pseudorandom binary sequence (PRBS). It is easily generated by means of a feedback shift register, realized in hardware or software (Brignell and Buttle 1968, Buttle 1969). The PRBS is a periodic function that has the property of being uncorrelated over its period. This means that the autocorrelation function is a sequence of pulses separated by a time-lag equal to the period length. What makes it such a powerful form of stimulus is embodied in equation (2.29), which tells us that cross-correlating the stimulus and response signals allows us to recover the impulse response of the system, and hence by Fourier Transformation (with some correction) the frequency response.

An important proviso is that all responses of the target system must be over within the periodic time of the sequence.

This observation may prompt the question 'If you want the impulse response, why not simply apply an impulse?' There are a number of reasons why one might not wish to do this. What the PRBS does is to spread the power of the impulse out along the time axis.

If we are making measurements on a plant that is in operation a large impulse would disrupt that operation, whereas a PRBS stimulus, say switching on and off tiny increments of flow of a reagent, could be made subliminal. It is only the averaging properties of the cross-correlation process that eliminate the noise and all other uncorrelated signals, allowing the impulse response to be recovered from an output in which the effects of the extra stimulus are otherwise unobservable.

There are many other cases where the impulse stimulus is undesirable. An example is in time-of-flight measurements in liquids to identify charge carriers by their mobility (Hewish 1975). Releasing large amounts of charge can produce electrohydrodynamic instability, whereas small pulses of charge released according to a PRBS can have a negligible effect yet again allow the response to be recovered. The argument applies to many cases where large peak powers are to be avoided. The fundamental principle is that the PRBS is a stationary signal (§ 2.1.8) whereas the impulse represents a gross non-stationarity.

The second example concerns target systems that exhibit a leading zero. A simple example is the capacitor

$$I(s) = s\,C\,V(s) \ . \tag{6.49}$$

There are cases in which the information we require (eg the presence of polar species) is in the complex permittivity of the dielectric. In this case the capacitance is a function of complex frequency:

$$I(s) = s\,\varepsilon_o\,\varepsilon_r(s)\,V(s)\,A/d \ . \tag{6.50}$$

Now some polar species exhibit a very long relaxation time, which means that measurement by sweeping through sinusoidal frequencies of low magnitude can take forever, because the system has to equilibrate at each frequency. Thus a desirable alternative is the step function, whose Laplace Transform is $V(s) = V\,(1/s)$ (equation (2.10)).

Hence

$$I(s) = V\,(\,\varepsilon_o\,A/d\,)\,\varepsilon_r(s) \tag{6.51}$$

which means that the current waveform accurately represents the time response of the dielectric. Unfortunately, the non-polar responses of the dielectric (e.g. electronic) are virtually instantaneous, so there is an impulse of current at the origin of the step. In practice this will overload any electrometer used to measure the current, making the measurement impossible. Early workers tried to overcome this problem by using a relay to short out the first surge, an expedient of arresting crudity.

An obvious approach might be to apply a low-pass filter to the step before applying it. Here we come up against the problem of realization. The perfect low-pass filter is a block function in the frequency domain, and it is fundamentally unrealizable. However elaborate the realization there will be substantial errors in the representation of the block function (figure 6.10) both in amplitude and phase. As a result the information of complex permittivity will be unacceptably distorted.

There is a solution to this problem that is a very nice example of the radical redirection of thinking that can come with the provision of digital processing, and exemplifies the intelligent sensor approach. It is reached by posing the question 'If we had a perfect filter what would the output be for a step input?' In fact this is very easily calculated. Consider the differentiated form of the step, which is the impulse, $\delta(t)$, with the Fourier Transform $F(j\omega) = 1$. We can represent the process of perfect low-pass filtering with cut-off frequency $\omega_c$ by changing the limits of the inverse Fourier Transform.

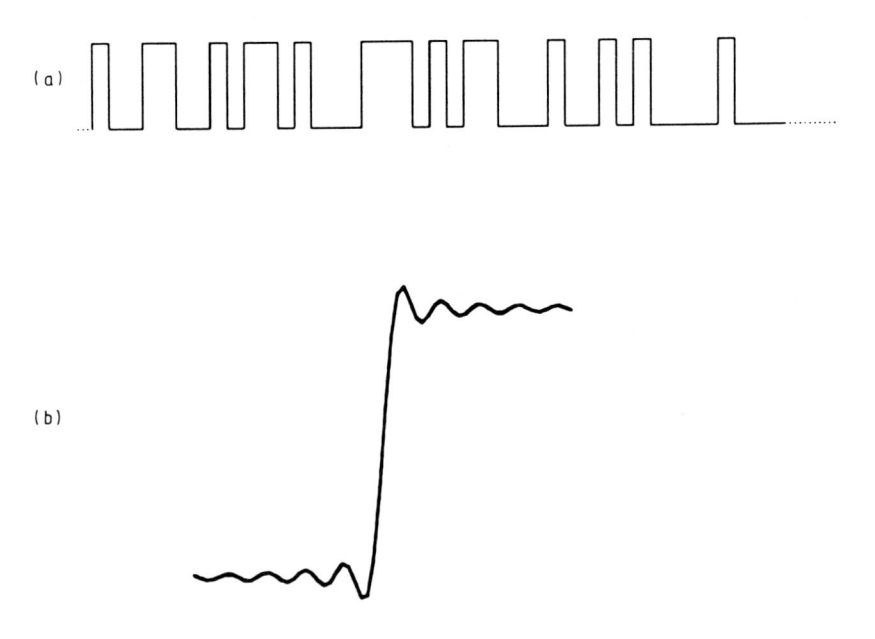

**Figure 6.21** Unorthodox stimulus waveforms (a) part of a PRBS signal of length 127 and (b) the *Si* function test signal.

$$f(t) = \frac{1}{2\pi} \int_{-\omega_c}^{+\omega_c} 1 \cdot \exp(j\omega t) \, d\omega = \frac{\omega_c}{\pi} \, sinc \, (\omega_c \, t) \, . \qquad (6.52)$$

This result (cf figure 6.14) shows why the perfect filter is unrealizable. It is a **non-causal** system, since half the response occurs before the stimulus. In simplistic terms the perfect filter is a system that knows what is going to happen before it does. We cannot create a perfect filter, but we can create the result of perfect filtering of an impulse, which is a *sinc* function, and by integrating this we can create the perfectly filtered step, which is expressed in terms of a transcendental function, *Si(x)*, the integral of *sinc(x)* (Flugge 1954).

Figure 6.21(b) shows such a waveform. It can be calculated by means of well known series, or more easily stored in a Look Up Table. It must be noted that in using this function we have to have a starting point, which is conveniently chosen as the left hand point at which the oscillations are less than the quantization level. Note also that in doing this we have shifted the step in the time axis, but fortunately this is easily corrected by taking advantage of the cyclic nature of the discrete transform (§ 6.4.3) and performing a rotation. By this means accurate measurement of the complex permittivity of polar materials with long time constants have been made (Chowdhry *et al* 1988).

The two examples of unorthodox stimulus functions shown in Figure 6.21 illustrate well the availability of powerful methods that can be used in stimulus-response sensors. We shall examine one simpler in concept later in this text ( § 8.3).

## 6.7 DISCUSSION

This chapter has been a key one. The treatment of intelligent sensor concepts has ranged from some of the basic circuit configurations to the broader issues of how the approach can be adopted in applications as various as buildings and automobiles, and we have discussed the vital issue of how the concepts affect production and its costs. Finally we have identified just one of the areas where the possibilities of the technology have barely been touched. This diversity of material underlines the marks at the beginning of this text as to the multi-disciplinary nature of the subject and the reminder that a little learning is a dangerous thing.

# 7

# Communication and Sensor Networks

## 7.1 SYSTEM TOPOLOGIES

One of the most important considerations in multisensor systems is that of the topology of the network connecting the sensors to the central processor. The classical arrangement is the star topology, shown in figure 7.1 (a). Here each sensor has at least a pair of wires connecting it to the centre. The disadvantages are fairly obvious from the topological diagram. Firstly, there is a great deal of cabling, and in a large industrial system this can become the dominating contributor to the system cost. Secondly, there is a bottleneck at the centre where all the cables arrive. This can be a major engineering problem, especially as modern industrial measurement systems tend to grow with time, and the re-engineering of the juncture can be expensive.

The modern alternative is the bus topology, which is shown in figure 7.1 (b). In this the sensors and actuators share a common pair of wires. Again certain disadvantages are obvious. First, the devices have to be addressable, which means that they have to have a certain amount of in-built intelligence. Second, if the bus is severed at any point all the devices beyond that point are disconnected from the system, which can be catastrophic in certain circumstances. Third, the devices have to share a common resource, the bus, and as their number increases their share of the resource, under time division multiplexing, diminishes. This third apparent disadvantage is not quite as dramatic as it seems, since even in the star connection devices have to share resources such as input ports and processor time.

As stated above the question of cabling costs is dramatically important; so much so that in star topologies it is common to adhere to a protocol that allows power to be delivered to the devices along the same pair of wires that carry the signal (§ 7.3.3.2). This can be very restrictive on system design, and while there is a tendency for traditional users to expect the same facility in bus topologies, it is generally not worthwhile to impose such constraints on these. A fairly naive calculation shows the advantage commanded by a bus system where there are many devices connected together. Imagine that all the devices, say $N$ in number, are placed round a circle of radius $R$. A star system would require a total cable length of $NR$, whereas they

INTELLIGENT SENSOR SYSTEMS                                      205

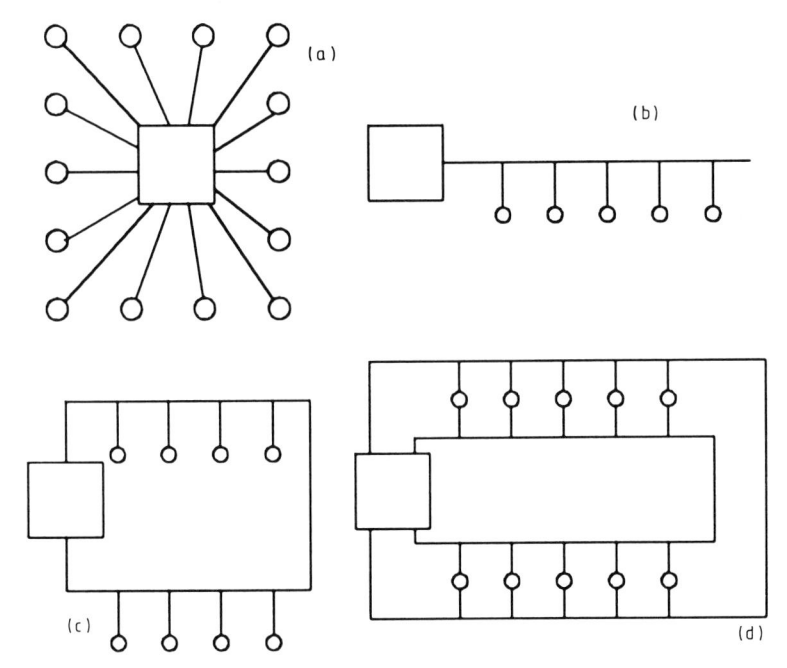

**Figure 7.1** Examples of network topologies; (a) star, (b) bus, (c) ring, (d) double looped ring.

could be completely connected by a bus of length slightly less that $2\pi R$ (actually $2NR \sin(\pi/N)$ if they are regularly spaced). Thus the ratio of star cable length to bus length is $N/2\pi$, which illustrates the economic dominance of cabling in star systems.

The vulnerability of the simple bus system to cable severance can be overcome by the modification illustrated in figure 7.1 (c). Here the bus is arranged in a complete loop, and provision is made for it to be driven from either end. This has a double advantage. If the cable is severed at one point the system can carry on by addressing both ends separately. It can also, however, locate the position of the fault by determining which devices fail to respond from either end. In cases where there is a particular danger of disruption, for example where there is an explosion hazard, then a belt and braces approach can be adopted, as shown in figure 7.1 (d). In this a double looped bus is addressed by four separate drivers. The bus separation, and hence the length of the stubs connecting the devices, is made large enough to minimize the probability of both buses being disrupted by a catastrophic event.

### 7.1.1 Prioritization

There is a consideration which exists whatever the topology used, that is the question of the relative importance of the connected devices as far as system response is concerned. In the extreme case there may be a mixture of very demanding devices with ones of low demand, in which case it may be necessary to provided separate buses or special interrupt systems. Otherwise the matter has to be dealt with in the way that the devices are polled, which is generally largely a software exercise. An outline of a simplistic software system is illustrated in the flow diagram of figure 7.2, which can be switched between a prioritized and an unprioritized mode. There are various possible possible levels of sophistication.

It would be dangerous, however, to understate the problem of implementing a prioritization scheme. It must always be a compromise, and increased access to the communications system for one device must always mean decreased access for another. The random nature of demand means that the average and maximum response times for any device is the function of the demands of all the other devices. We must remember that the maximum information rate carried by the binary channel is determined by the transmission rate and the source entropy (§ 1.2). Virtually every addition we make to the transmission protocol, e.g. dividing the message into fields, reduces the efficiency of coding and hence the rate of transmission of information.

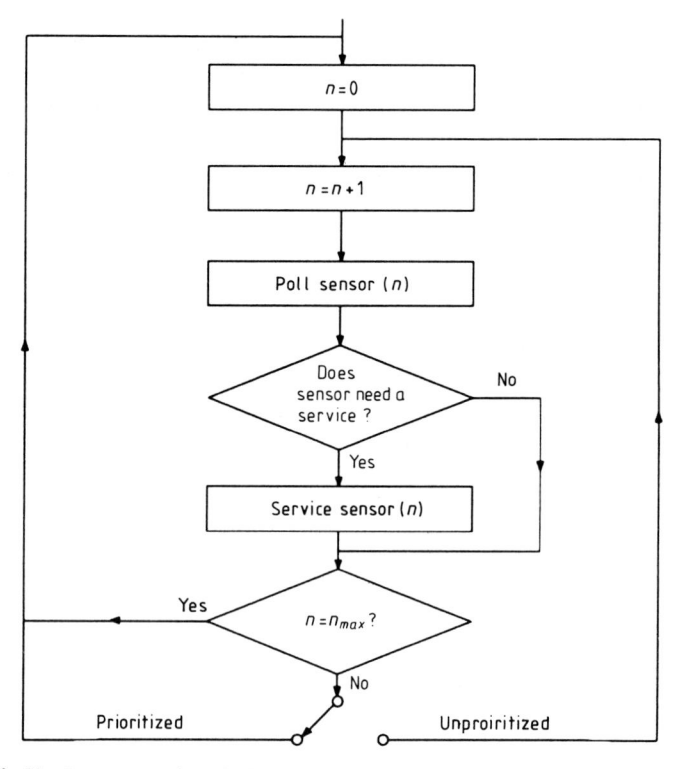

**Figure 7.2** Simple representation of prioritized and unprioritized polling routines.

## 7.2 GENERAL REQUIREMENTS OF A LOW-LEVEL PROTOCOL

It is a simple matter to determine the facilities which would need to be provided for messages to be sent and received on a shared bus, and this can be approached by considering what questions would be left open in the complete absence of any protocol. Imagine the situation in which a continuous stream of bits is being received by a station on a bus. Obvious questions are posed:

Where does a message begin and end?
Is the message for me or some other station?
What is the purpose of the message and how is it formatted?
What is the actual information contained in the message?
How can I ensure that the message has been properly transmitted?

These questions immediately indicate a minimal number of fields that are required to establish a workable protocol. As in all engineering design the actual definition of the protocol is the result of a number of trade-offs, and the different protocols are the result of different approaches to such trade-offs, though it must be stated that the existence of so many competing definitions is more a political than a technical problem.

A very fundamental trade-off in this design process is between information flow rate and information integrity. Our communication channel will have a certain capacity in bits/sec, determined largely by the bandwidth, noise and other mechanisms for signal degradation. As we have seen in § 1.2, the moment we start dividing our stream of bits into fields we begin to reduce the coding efficiency from $r$ the maximum prescribed by the **source entropy**, and as we introduce checks on

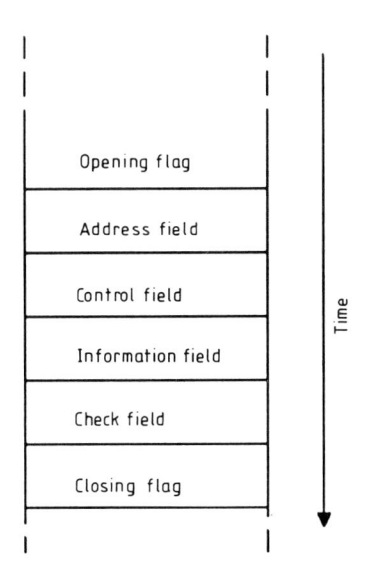

**Figure 7.3** A general low-level protocol.

the integrity of information by means of **redundancy**, so we progressively reduce the efficiency of **coding**, and hence the rate at which information can be transmitted.

Figure 7.3 illustrates the division of our bit stream into fields in a basic minimal protocol. Firstly, we require an opening flag. This has to be a unique signal which cannot occur by accident anywhere else in a message. This is a non-trivial problem, and its solution is one of the chief characteristics of any given protocol. There are obviously solutions of a hardware nature, such as having a pulse of a different height, width or polarity from the pulses representing the message bits. Alternatively there are solutions which preserve the uniqueness of the flag by proscribing it from the rest of the message bit-stream, again at the expense of source entropy.

The next field required is the address field, and with a shared bus it is clearly essential that the destination of any message is accurately and unambiguously defined. After this comes the control field, which is a much more variable item. Basically it is there to make one or more statements about the purpose and nature of the message. There will be a list of prescribed codes with which it conveys such instructions to the receiving station. Typically they might be might be one or more of the following:

The following field is of 32 bits divided into four octal sub-fields.
Carry out a self-test and calibration.
The following field is a new setting for the input amplifier gain.
Send values of temperature and pressure.
Confirm that the last pressure sent was that quoted in the following field.
etc.

The next field in this general model is the all important information field. It is vital to note that its interpretation is highly conditioned by the fields that have gone before. In particular the list to decode the control message may be unique to the particular address or group of addresses. Ideally the information field is of variable length, but some protocols adopt fixed quanta, which may mean that multiple messages may have to be sent in the case of a piece of information exceeding the number of bits prescribed.

The penultimate field is the check field, which in general is a number derived from the preceding sequence of bits by a process which is reproduced in the receiver and checked for a match. In the event of the result being discrepant a request is issued for the information to be transmitted. Finally there is another transmission of the unique flag to indicate the end of the packet, and perhaps the beginning of the next one.

In the present treatment we do not propose to go into the detail of these low level protocols, but it is useful to look at one example to illustrate how these general principles can be illustrated. The well known protocol HDLC is illustrated in figure 7.4. Its unique characteristic is the method of defining the flag. The flag is a binary pattern of the form 01111110, and in order to preserve its uniqueness, a technique known as bit- stuffing is applied to the all the non-flag section of the message. The bit stuffing mechanism adds in an extra zero at the transmitter whenever a sequence of five consecutive ones occurs. At the receiver, after a sequence of five ones a

| Flag | Address field | Control field | Information field | Frame check sequence | Flag |
|---|---|---|---|---|---|
| 01111110 | 8 Bits | 8 Bits | Variable | 16 Bits | 01111110 |

**Figure 7.4** The HDLC protocol.

following zero is removed, while a following 1 must represent a flag pattern. An advantage of HDLC is that the information field is of variable length, and while it is argued that bit stuffing does not give the best security of transmission, in practice this is found to be a reliable protocol well supported by dedicated silicon devices.

## 7.3 PROTOCOL IMPLEMENTATIONS

### 7.3.1 Communication systems reference model

If we consider a network of interconnected sensor systems then, as mentioned earlier, we need a method of initiating and maintaining communication between each device connected to that network. In view of the vast number of possible protocol implementations and the need for compatibility or 'openness' between different manufacturers equipment, the International Standards Organization has defined a seven layer communication protocol which allows an Open Systems Interconnect (OSI) strategy. This reference model is shown in figure 7.5.

The idea behind the layering is that each layer adds to the services of the layer(s) below (OSI/ISO reference model 1983). If an error occurs and is detected on one level then a request is sent to the transmitter of the data to re-transmit. Only when no errors are detected will the data be passed on to the next layer above. The following gives a brief description of the functions and services provided in each of the seven layers.

*Physical layer.* This is the lowest layer of the model and is concerned with the interface to the interconnecting medium. Standards defined include the types of connector to be used, the electrical levels of the signals on the data bus and other procedural aspects such as handshaking. The reference model does not include a

| | |
|---|---|
| Application | 7 |
| Presentation | 6 |
| Session | 5 |
| Transport | 4 |
| Network | 3 |
| Data link | 2 |
| Physical | 1 |

Physical medium

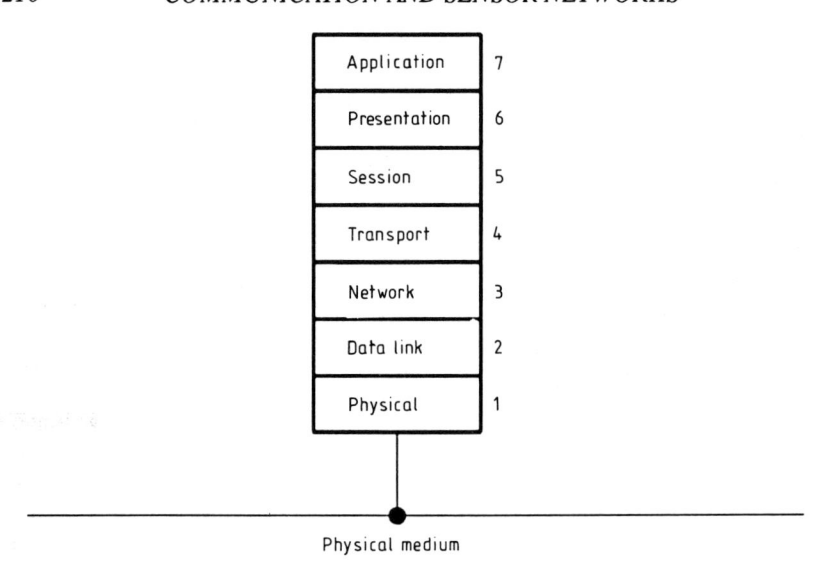

**Figure 7.5** Seven layer OSI/ISO communications reference model.

specification for the physical medium itself and hence the type of cable (twisted pair, co-axial etc.) is not defined.

*Data link layer.* This layer is responsible for maintaining reliable transmission of data over the network. It is concerned with important aspects like framing and the detection and correction of errors which may have occurred in the physical layer.

*Network layer.* The main function of this layer is to mask certain characteristics of the transfer medium from the layer above. For example, the transport layer does not require a knowledge of the origin of the data i.e. from a local area network, satellite link, telephone network etc. This layer sets up the appropriate communication paths for the messages.

*Transport layer.* Essentially this layer provides a transparent data transfer service between the communications technology employed and the user of the system. It allows processes to exchange data reliably.

*Session layer.* This layer is responsible for managing the interactions between several user applications on different systems which are connected via the network.

*Presentation layer.* The main function here is to ensure that the user application does not depend on any differences in the presentation of the data i.e. the syntax.

*Application layer.* The services provided on this layer may be called directly by the user. Examples here are file transfers, remote file access and terminal operation.

Clearly, in order to achieve apparent transparency between layers in the model a price must be paid, both in terms of the actual cost of the hardware/software involved and also in terms of the overall data transfer rate capability of the network.

### 7.3.1 Conventional instrumentation communication systems

#### 7.3.2.1 *General Purpose Instrumentation Bus (GPIB, IEEE488)*
The GPIB consists of sixteen wires, eight parallel data lines and eight control lines. It is often used in laboratories for automatic control of equipment or data acquisition. The instrument bus allows for the interconnecting of oscilloscopes, voltmeters, microcomputers and the like, over relatively short distances (20m or so). The GPIB provides for a high speed transfer of parallel data (up to 1 megabyte/sec), with full handshaking among up to fifteen devices. One device, normally the host computer, has the task of assigning other devices to be talkers or listeners with the restriction that only one can talk at a time.

Generally, GPIB is regarded as being too expensive to implement for the interfacing of transducers for use outside the laboratory. GPIB interface circuits are readily available and relatively low-cost but the major factor which inhibits the use of parallel data transmission in industrial instrumentation systems is that of the associated wiring costs.

#### 7.3.2.2 *4-20 mA Current Loop*
This is probably the most common analogue communication system currently in use within the process control industry. Figure 7.6 shows a schematic for a two-wire system.

The supply voltage is typically in the range 10 V to 30 V and is used to supply power to the interface circuit and subsequently the sensor itself. There are only two external wires connected to the the transmitter. The limited use of cabling is

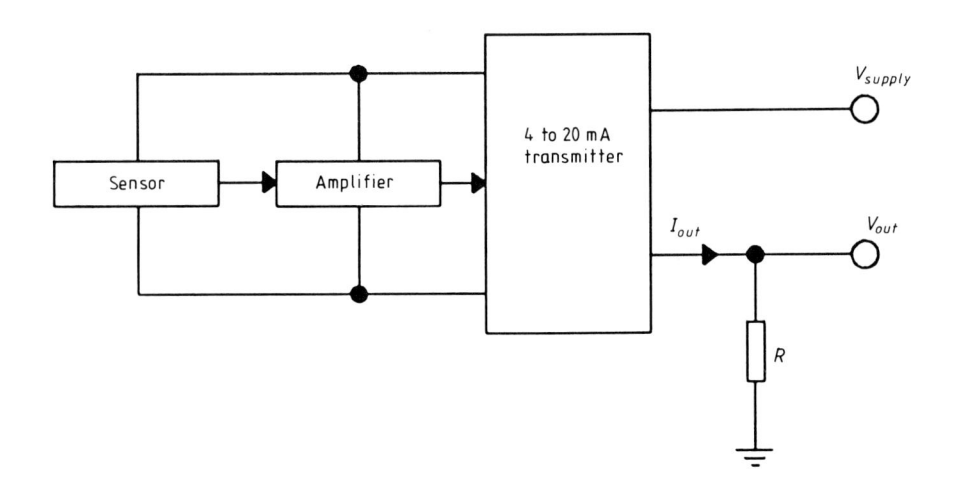

**Figure 7.6** Schematic diagram of a 4-20 mA transmitter.

one of the attractions of this system. The output current, $I_{out}$, is constrained between 4 and 20 mA with 4 mA representing zero and 20 mA corresponding to full scale. The use of current as a signal has an additional advantage, in terms of noise immunity, over voltage-derived signals. The technique of using the 4 mA as a live zero also allows failure to be detected. For example, an open-circuit condition would result in zero output current being measured and a short-circuit fault would be seen as an output current exceeding the 20 mA limit. It is usual to connect an external reference resistor, $R$, between $I_{out}$ and ground so that the recipient actually measures a voltage signal (usually in the range 0.4 to 2 V).

There are also a number of disadvantages associated with a current loop communications system: limited diagnostic information, one way communications ability, no direct compatibility with the host computer and poor quality of information. The need for a digital communications standard thus becomes evident.

### 7.3.3 Manufacturing Automation Protocol (MAP)

MAP (MAP reference specification 1988) is a communications protocol based on the full seven layer ISO/OSI reference model. The American automobile manufacturer General Motors were largely responsible for its development. In the early 1980s GM realized that their manufacturing capabilities needed to be automated in order to remain competitive. Hence a standard communications system was proposed which would allow compatibility between existing items of equipment.

The physical and data link layers of the protocol are based on an ISO/IEEE broadband token bus LAN standard. This allows several carrier band signals, at different carrier frequencies, to be combined onto a single physical transport medium. Various digital modulation techniques are used to generate the carrier band signals. Examples of such methods include amplitude shift keying (ASK), frequency shift keying (FSK) and phase shift keying (PSK). The major advantage of using a broadband technique is that the data highway can carry additional data such as telephone, closed circuit television and other computer data. In order to provide a sufficient bandwidth for each channel the carrier frequencies are typically in the VHF range (200 MHz).

As MAP uses the full seven layer protocol stack the requirements for the hardware/software processing are very severe and so MAP is not widely used as a general purpose instrumentation standard. One suggestion is that networks carrying transducer data can be connected to a broadband network via a gateway, the function of which is to convert the protocol of one network into a compatible form with another.

### 7.3.4 Enhanced Performance Architecture (EPA) MAP

The MAP/EPA specification provides for a system which uses a reduced protocol stack in which the layers between the data link layer and application layer are eliminated. At the physical layer the broadband technology is replaced with a carrier band technique so that only a single channel is available. Data are usually transmitted using a frequency shift keying (FSK) method where two different frequencies are used to represent a logical 1 and 0 respectively. This is much cheaper to implement than the broadband techniques.

The data link layer is essentially the same as that used in the normal MAP system except that the correct reception of data by the receiving system must be acknowledged at the data link layer before proceeding to the application layer. At the top level the application layer provides a functional interface between itself and the data link. Owing to the fact that there are several missing layers the uppermost level has to be capable of fulfilling many of the missing intermediate tasks.

Although MAP/EPA achieves a fast response time by reducing the protocol stack and has a lower implementation cost than MAP, the system is not strictly 'open'. In order to achieve open systems architecture connection to a suitable gateway interface required.

### 7.3.5 Fieldbus

Fieldbus is a name given to a communication system specifically designed for the networking of transducers. That is, fieldbus is intended for applications at the lowest levels in the manufacturing systems or process control hierarchy namely devices in the field (Bradley *et al* 1991). At the time of writing there is no accepted standard for fieldbus although the International Electrotechnical Commission (IEC) is responsible for defining one. The term FIELDBUS in upper case letters is reserved for the internationally agreed standard (Atkinson 1991). A three layer protocol model has been adopted for FIELDBUS and this comprises the physical layer, data link layer and application layer.

The definition of the physical layer has been in a higher state of maturity than that of the subsequent two layers. Two distinct physical layer definitions have emerged, one for low speed communications (31.25k bits/sec) and the other a higher speed version (1M bits/sec). This resulted from an initiative put forward by the Instrument Society of America (ISA) which highlighted the need for low data rate, long distance communications (H1) and high data rate, short distance communications. Both the high speed and the low speed implementations will use a voltage mode of operation with direct connection to the bus. Another option is for a current mode version which is intended to be aimed at intrinsically safe applications. It is estimated that such an implementation would be transformer coupled to the bus.

The definition of the data link layer is a stage further on. One idea is for a token bus principle which offers the advantage that integrated circuits are readily available which already perform this task. One area of concern here is the manner and timing of the token rotation which needs to be modified to give deterministic behaviour.

The application layer has to make the FIELDBUS network appear transparent to the application program in each transducer connected to the bus. At the time of writing there is no agreed standard for this very important top level protocol.

### 7.3.6 The HART communication protocol

HART is a communication protocol developed by Rosemount Inc. and is an acronym for Highway Addressable Remote Transducer Protocol. Hart is a digital system but it preserves the integrity of the 4-20 mA current loop signal (Vincent 1991). Like most of the sensor communications systems already discussed, HART is based on the reduced stack OSI/ISO reference model. The layers used are the physical layer, data link layer and the applications layer.

The physical layer utilizes the same hardware as the 4-20 mA convention. A high frequency FSK current signal is effectively superimposed on the analogue signal. The two frequencies used are 1200 Hz and 2200 Hz which represent the binary values 1 and 0 respectively. The mean value of the digital signal is zero and therefore no further DC. component is added to the existing analogue signal. This provides the attraction of allowing compatibility with conventional 4-20 mA devices.

The HART protocol uses a master/slave system where the master devices transmit a voltage signal and the slaves transmit a current signal. The current is converted to a voltage via the loop load resistance as described earlier. Hence any device connected to the loop can use a voltage sensitive receiver circuit. The maximum allowable cable length for this system is 1500 m but a reduced distance may need to be implemented if poor quality cabling is used.

The data link layer performs the functions outlined in the earlier sections on three layer protocols; i.e. this protocol detects for errors generated on the physical link medium and requests retransmission of the data when corruption is detected. As HART is typical master/slave protocol the master is the only device which may initiate a message transaction. Each slave device has a unique address and will only respond when it receives a message containing its own address. Information transfer between devices on the network is accomplished through a defined message frame format like that of figure 7.3

At the application layer a number of message types are defined. The three groups are: universal messages, common practice messages and devices specific messages. An example of a universal command would be reading the direct measured value from the sensor. The idea of having common practice commands is that a set of commands can be implemented across a number of similar device types and would include calls for setting the span of the device and initiating self-testing routines. Different sensors may well need some device specific commands issued to them. For example, the calibration procedure of a pressure sensor will be different from that needed for a temperature sensor.

A number of companies have adopted HART as their sensor communication standard as it offers the benefits of digital communications together with their existing analogue systems. In the long term, if the full FIELDBUS standards are universally adopted, it is envisaged that HART will almost become redundant.

# 8
# Physical Realizations

## 8.0 INTRODUCTION

In this chapter we attempt to draw together some of the threads from the earlier sections. Four apparently disparate case studies reveal the common themes developed earlier. In each of them the different classes of compensation identified in § 3.2 are realized in quite different technologies, yet adhere to the basic principles established there. The elements of intelligent sensor systems identified in § 6.1 are also present, but again in various guises. Because of exigencies of space we have not given a complete description of each structure, but have rather emphasized aspects which illustrate this variety of realizations of these fundamental principles. Indeed, the examples have been selected on the basis of how well they illustrate those principles rather than as representatives of the state of the art. Also we have avoided illustrations where the advantages of particular processing techniques are fairly obvious, spectral analysis for example.

The first case, discussed in § 8.1, is a magnetic field sensor. This was by way of a pilot study into the problems of designing an intelligent sensor, carried out in the authors' laboratories at the beginning of the 1980s, and as such was perhaps one of the earliest realizations. Although now somewhat obsolescent it furnishes a valuable didactic illustration of the basic hardware and software sub-systems that go to make up a working device.

The second example, in § 8.2, is the intelligent load cell. Much of what is involved here has been adumbrated in earlier paragraphs, as the problems posed have been valuable as general examples. The load cell exemplifies our earlier statement that the great majority of mechanical sensors are fundamentally second order systems (§ 2.2.6), and a prime requirement is to eliminate the spurious response that this observation entails.

Interestingly, the third example also evinces this form of response, but in this case we seek to exploit it rather than eliminate it. The technology here, in § 8.3, is quite different, in that fibre optic techniques are used to create a pressure sensor that combines safety critical features with immunity from electromagnetic interference.

The fourth example is in many ways a triumph for the principles of intelligent sensors, the electronic nose, treated in § 8.4, in that it shows how an array of sensors, each individually so unspecific as to be virtually useless can, be combined to produce a gestalt of immense potential power and application.

In § 8.5 we turn away from application examples to look at the way the new dominant technology (ASICs) affects our approach to the design problems. Finally, in § 8.6, we draw a few brief conclusions from our voyage from the most fundamental and simple of principles, via the basic building blocks, to the more detailed realizations.

Particular note should be taken of how these sensors can be trimmed ( or in our vocabulary given tailored compensation ). It is a process of downloading coefficients into ROM within the intelligent sensor following a test procedure that can be automated. The procedure was summarized in figure 6.20, which show sensors coming along a production line, having an automatic test applied, and then receiving the requisite information as a set of numbers which represent the full compensation procedure. It should be re-emphasized that the most difficult part of this process is often the control of the target variable in the post-production test.

## 8.1 A MAGNETIC FIELD SENSOR

### 8.1.0 Introduction

In 1980 the idea of intelligent sensors was still somewhat controversial in some quarters, and a project was initiated in the authors' laboratories to try to determine just what would be involved in the design of such devices (refs). For such a pilot study a requirement was identified to combine the important enabling technologies – custom chip design, microprocessor hardware and software, thick-film hybrid interconnection, packaging and serial data communications. In order to obviate unnecessary complications, such as surface effects, it was decided that a bulk effect in silicon would be exploited, and the magnetic Hall effect was an obvious choice. The objective was to produce a complete transducer for magnetic flux that would be fully compensated internally for linearity and temperature cross- sensitivity, and would communicate with a host system by being addressable on a multidrop serial bus.

### 8.1.1 Mechanisms

The Hall effect having been chosen, there remains the choice of the actual structure. The magnetotransistor (§ 3.3.4.3) is an obvious choice for incorporation into a custom chip, as the disposability of the drain loads aids the flexibility of design and the structure lends itself to inclusion in a fairly conventional MOS layout. At the time this work was done AD conversion was not available as a chip sub-system, so voltage to frequency conversion was chosen as a mechanism. This is not entirely disadvantageous as frequency encoding is a very good way of getting information into microcontrollers, since they tend to have independent time-counters that can operate without impeding the programmatic operation.

### 8.1.2 Structural compensation

In choosing the structure of the custom chip we appeal to our most fundamental version of structural compensation, which is design symmetry (§ 3.2). Each of the elements at our disposal (the magnetotransistor, amplifiers and voltage controlled oscillators) is temperature dependent. The magnetotransistor itself (§ 3.3.4.3) is symmetrical, so by duplicating amplifiers and VCOs we can create a structure that is completely symmetrical. This can be seen in figure 8.1. In theory, changes in temperature and other interfering variables and parameters affect each half equally and therefore do not appear in the differential output. In the real world, of course, no two entities are equal, and there is the further problem that this argument only deals with offsets and not gains (figure 3.4) so there is always a residual effect.

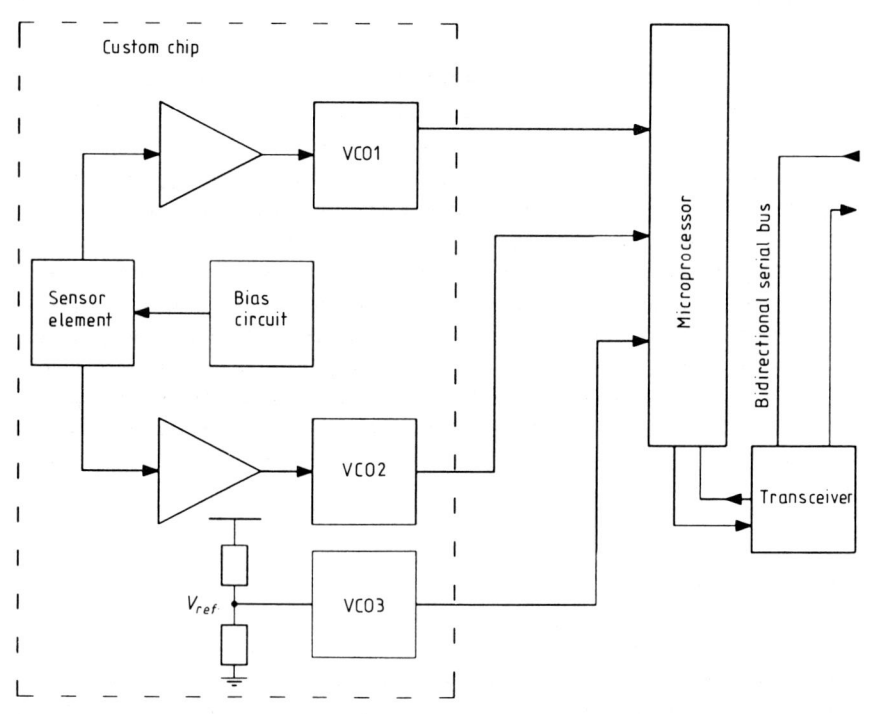

**Figure 8.1** Layout of the intelligent magnetic field sensor.

## 8.1.3 Monitored compensation

Temperature, as is usually the case, represents the dominant interfering parameter. The solution in this case is to turn an apparent disadvantage into an advantage, not an unusual tactic in intelligent sensor systems. The VCO is temperature dependent, so why not use it as a temperature sensor, with the added advantage that it conforms to the frequency encoding principle already employed? A third VCO is incorporated with a constant reference voltage applied, and this personates the vital monitoring line of figure 6.1.

## 8.1.4 Information processing

The essential element of information processing in this device is the implementation of a two-dimensional look up table. The output of the custom chip in figure 8.1 is largely a function of two variables – magnetic flux and temperature. The raw magnetic flux information is represented by a frequency difference, while the temperature is represented by a third frequency. Furthermore, the devices used, and in particular the VCOs, are non-linear. Nevertheless, for any combination of the two pieces of input information there is a unique true magnetic flux which can be accessed by a two dimensional LUT (§ 6.2.2.1).

## 8.1.5 Tailored compensation

Although this device was a research tool and not intended for production its manner of tailored compensation is very illustrative of how the intelligent sensor can be trimmed in the production environment (§ 6.5). In order to calibrate the sensor it is necessary to apply a number of known magnetic flux densities at a number of temperatures. It is not necessary for the device to know the temperature in degrees, it merely has to equilibrate at a number of temperatures which it can register in its own internal units (in this case frequency). By the initiation of an internal calibration program it fills in the entries to its internal LUT, as illustrated by the surface in figure 8.2. Thereafter its output is fully temperature compensated and linearized.

## 8.1.6 Discussion

Although early and somewhat primitive in its construction, this device fulfilled its role in identifying the characteristics and problems associated with an intelligent sensor design. Some of the most important principles, such as structural compensation by design symmetry and monitored compensation for temperature, are well exemplified. The physical realization of the device is shown in figure 8.3. It is a thick film hybrid circuit with one custom chip and a little glue logic but dominated by a microprocessor package, for at the time compactly packaged devices were

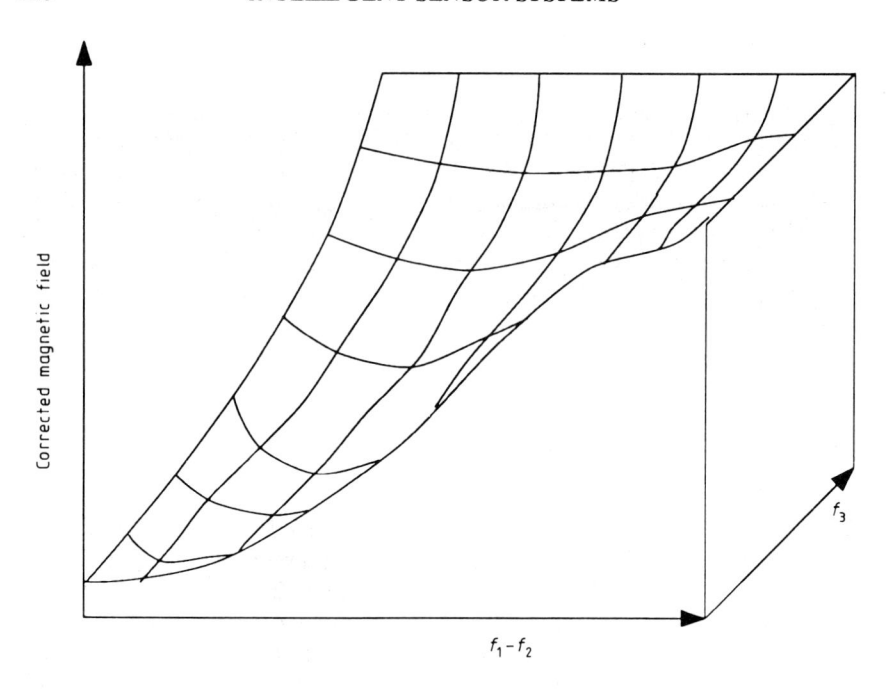

**Figure 8.2** Diagrammatic representation of the two dimensional LUT used in the program of the intelligent magnetic sensor.

unavailable. It will be noted that the microprocessor is of the type with UV erasable ROM.

The completed device also met the requirements of an intelligent sensor as laid out in the early part of this text. It responded to its own unique address on a serial bus via the HDLC protocol (§ 7.2). It reported magnetic field, with temperature and non-linearity errors reduced to negligible levels, and carried out a number of self-check routines automatically (Cooper and Brignell 1984, 1985).

## 8.2 THE INTELLIGENT LOAD CELL

### 8.2.0 Introduction

As observed in the introduction to this chapter, much of the material relevant to the intelligent load cell has been foreshadowed earlier in this book. This is because the process of weighing exemplifies one of the most significant complications of measurement, the situation where the measurand (in this case load) is also a significant parameter of the overall system. As a result the weighing system is

# PHYSICAL REALIZATIONS

**Figure 8.3** Photograph of the intelligent magnetic field sensor.

non-linear (i.e. non-additive and inhomogeneous, § 2.1.2). This is a good example of a problem that is totally intractable in traditional sensor technology, apart from the undesirable cure-all of applying massive damping. Because the parameters of the sensor system are changing, it is necessary for the roots of any compensating filter to track them accurately. It is perfectly possible to control the parameters of a continuous filter by using voltage controlled resistors, e.g. in the form of a FET. What is not possible in continuous systems, however, is to store and implement a control law in the form of a complex locus as we have successfully demonstrated in § 6.4.3. As we saw there the intelligent sensor approach to this problem is to try to cascade two non-linear processes to produce a quasi-linear one. In many weighing systems response time is important. In human controlled weighing waiting for the system to settle is not tolerable, while in automatic weighing, such as on a conveyer belt, slowing down the process to allow for load cell settling time could be dramatically expensive. Furthermore recent legislation by bodies such as the EC has made rapid and accurate weighing of packages coming off the production line an imperative.

Apart from response time, the other consideration in weighing is drift. Load cells as primary sensors have, over recent years, been developed to a high degree, and the structural compensation is now highly effective.

## 8.2.1 Mechanisms

The load cell provides a good illustration of cascaded sensor blocks as set out in figure 3.2. The relevant version is shown in figure 8.4. The metal structure converts the variables of force and displacement to stress and strain, while a strain gauge bridge in turn converts these to the electrical pair of voltage and current. The strain gauge bridge is a relatively simple sub-system, but the metal structure is more complicated than it might seem at first sight. It exhibits significant spring and inertial coefficients, so it is at least a second order system. Furthermore the basic 'impedance' ratio of stress to strain, Young's modulus, is a temperature dependent quantity. It is not always realized that this latter constraint provides a fundamental limitation to what can be achieved by traditional compensation methods.

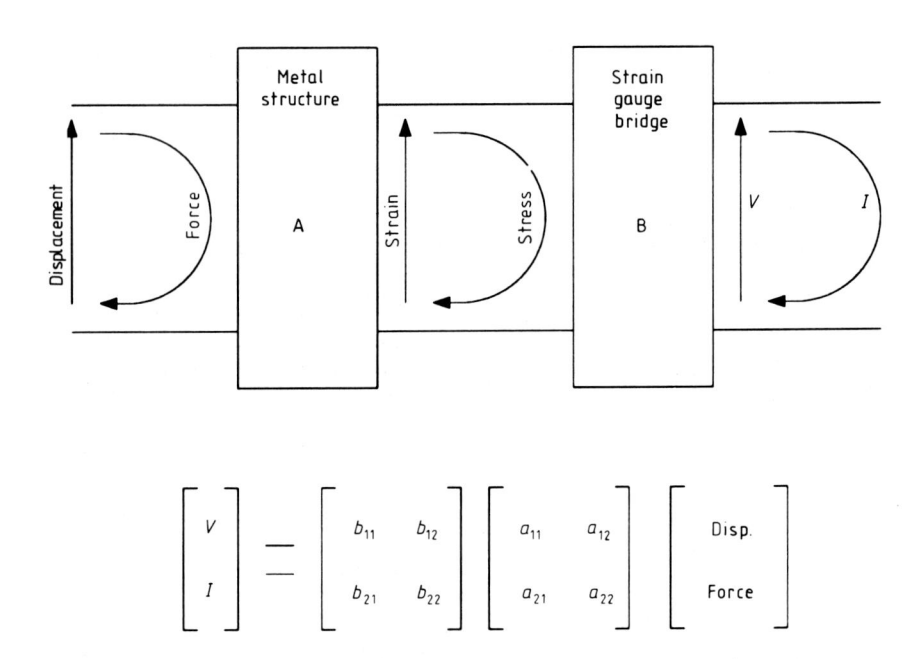

$$\begin{bmatrix} V \\ I \end{bmatrix} = \begin{bmatrix} b_{11} & b_{12} \\ b_{21} & b_{22} \end{bmatrix} \begin{bmatrix} a_{11} & a_{12} \\ a_{21} & a_{22} \end{bmatrix} \begin{bmatrix} Disp. \\ Force \end{bmatrix}$$

**Figure 8.4** The conceptual sub-system blocks in a load cell.

### 8.2.2 Structural compensation

The precision load cell is a classical example of the refinement of structural compensation (§ 3.2). Figure 8.5 shows in diagrammatic form the mechanical and electrical layout of a typical load cell. In both cases the symmetry of design stands out. This form of structure ensures that interfering variables and parameters appear as common mode signals. In contrast to a single gauge the full strain gauge bridge provides a maximum differential output with linearity (§ 4.1.1). The mechanical structure is in the form of a coupled beam (cf. figure 3.17), which offers a considerable degree of immunity from errors due to load eccentricity.

*PHYSICAL REALIZATIONS* 223

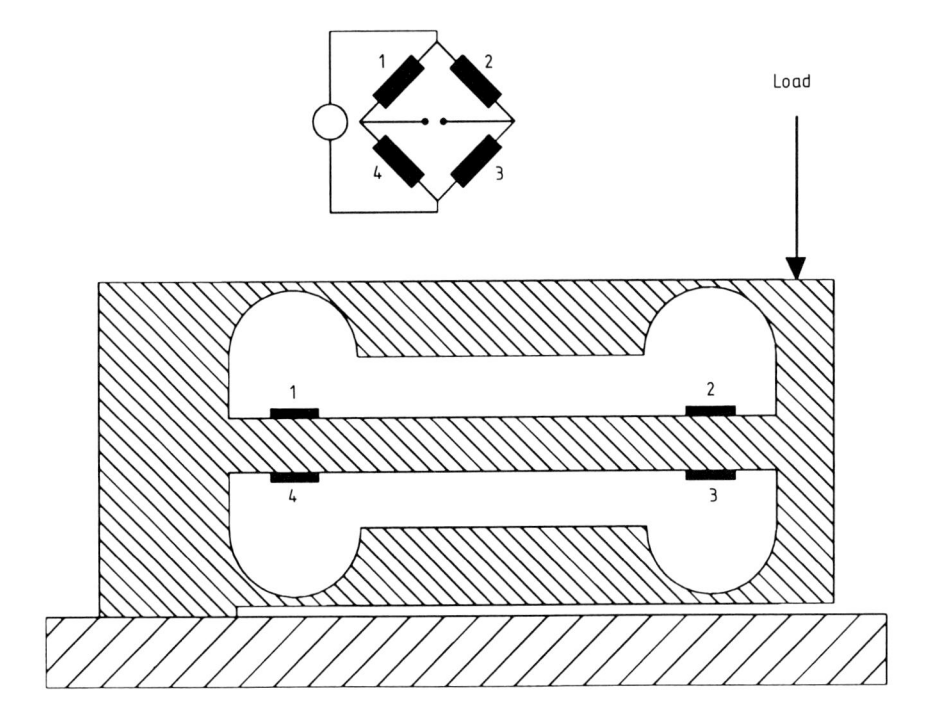

**Figure 8.5** Diagrammatic representation of the electrical and mechanical structure of a precision load cell, illustrating structural compensation by design symmetry.

In traditional manufacture this structural compensation was supported by tailored compensation. An eccentric load was applied in the form of a ball weight on a horizontal beam. This was rotated and small areas of metal were manually filed off the thin parts of the structure to optimize the immunity to eccentricity. Latterly this was done by robots. Further, each cell was cycled through its operational temperature range and the offset monitored. As a result a length of resistance wire of given temperature coefficient was calculated and this was added into one arm of the bridge. This is, of course, only a first order correction and gives only the best straight line fit to the temperature error characteristic. It should be noted that, as a residual error function, the form of this characteristic is quite arbitrary.

As we have observed it is not always appreciated that there is a fundamental minimum of temperature error due to the temperature coefficient of Young's modulus. This is an error of gain rather than offset (figure 3.4) and cannot be dealt with by the above simplistic approach.

In this particular study it is salutary to re-examine the primary sensor in the light of the availability of intelligent sensor methods. The classical load cell is an expensive item, being precision machined from a billet and subject to an equally expensive trimming process. The authors developed an alternative structure based on thick-film technology, therefore offering all its advantages of cheapness and robustness (§ 5.4.1). The structure is planar and produced by the inexpensive process of stamping out a steel shape and then printing it with gauges and conductors. The cell comprises three beams joined and supported at the centre with three pins near their extremities supporting the weigh pan. It can be shown theoretically that if the gauges at each position on each beam are connected in series the errors tend to average out and furthermore load eccentricity is balanced out, provided the centre of load is anywhere inside the circle passing through the three support pins.

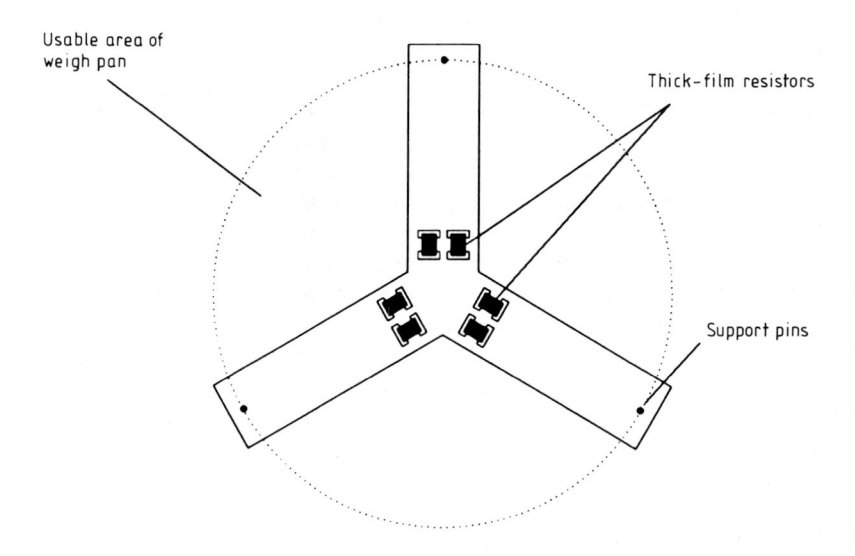

**Figure 8.6** Diagram of the tri-beam load cell. The planar cell is supported at the centre and pins on the arms support the weigh pan.

It interesting to note that this alternative structure (the tri-beam load cell of figure 8.6 (White and Brignell 1991) becomes viable only with the introduction of intelligent sensor techniques, but as it reduces the cost of the primary sensor component by more than an order of magnitude this is a worth while development. It would not be easy to use in a traditional system as it is mechanically resonant and underdamped with low effective mass, so not only is it subject to oscillation but the non-linear effect of the mass of the object to be weighed is emphasized (§ 6.4.3).

### 8.2.3 Monitored compensation

As we have observed (§ 3.2) the intelligent sensor approach to temperature compensation is to measure the offending parameter and remove its effect. Thus in the intelligent load cell will be found a primary temperature sensor, which could be any of the types mentioned in § 3.3.2, but the PTAT device is reasonably cheap and simplifies subsequent processing (figure 3.24). Load eccentricity could also be monitored by the use of strain gauge rosettes. In the case of the tri-beam load cell temperature is more cheaply and effectively monitored by printing a sensor on the steel substrate.

### 8.2.4 Information processing

In weighing systems, as in most sensing systems, we are usually concerned with extracting the maximum performance for a minimum price. Performance involves both the precision and speed of weighing. This consideration brings the fundamental non-linearity of the weighing process (§ 6.4.3) into focus. The system is quasi-linear, in that if we are prepared to wait for the transient response to settle down the relationship between input and output is linear. Indeed even after application of our adaptive filter (§ 6.4.3.2) the system is still quasi-linear, but the time delay is more than an order of magnitude shorter.

The traditional approach to dealing with the spurious response is to introduce a dominant real pole by means of mechanical or electronic damping. This, however, barely mitigates the problem of the delay between applying the load and being able to estimate its value.

It will be evident that, if we apply adaptive filtering immediately after data conversion we have effectively cascaded two non-linear processes to produce a quasi-linear one; so we can introduce further processing stages as though we were dealing with a linear system.

It will be noted that our approach to noise suppression by adaptive filtering (§ 6.4.3.3) we have also adopted a non-linear approach. This has to be dealt with much more circumspectly than the former case. In our introductory discussion on noise (§ 3.2) we defined noise as any unwanted signal, then went on to observe that noise is therefore defined by the nature of the wanted signal. The assumption upon which the adaptive noise filter is based is that the load signal is a step function. This is by far the most common circumstance, but there are others, such as the controlled filling of packages, in which case the signal is a ramp.

Drift, also, is a perturbation that is defined by its relation to the wanted signal (§ 3.2). It is one of the most serious problems facing sensor systems designers. Often there is no way of distinguishing between drift and the target signal and this is a limiting factor on accuracy. In many weighing systems, however, we have an extra piece of knowledge about the signal that can be utilized. This is that the weigh pan is periodically empty. In such a circumstance a drift tracking routine can be implemented.

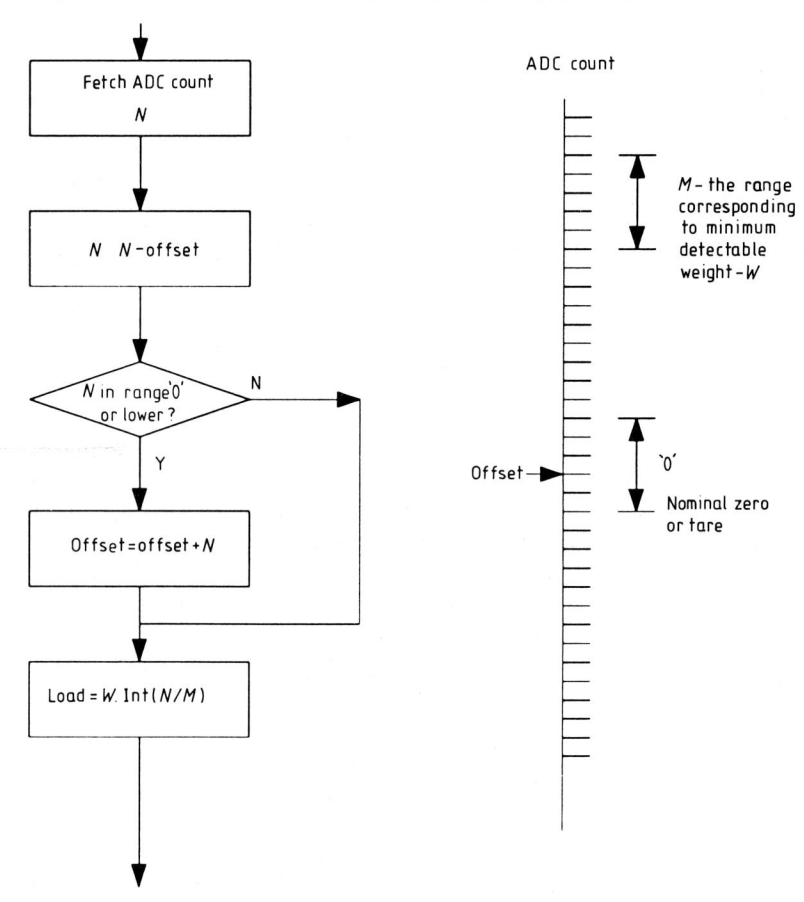

**Figure 8.7** A simplistic drift correction routine in the form of a flow diagram. It relies on the weigh pan being periodically empty.

One of the great advances that digital systems have facilitated in measurement is that variables can be stored indefinitely, a vital resource unavailable in older continuous systems. One such important variable is a stored offset. Figure 8.7 shows a program loop in the form of a flow diagram. An important requirement is that the precision of AD conversion be greater than the required precision of weighing. It should be noted that when offsets are likely to be large it important that they are dealt with before rather than after AD conversion, otherwise they are likely to swallow up the dynamic range of the ADC (§ 6.2.1.2, 6.2.1.4).

### 8.2.5 Discussion

The load cell has been a valuable example in this text. It shows very well the principles of structural compensation by design symmetry in both the electrical and mechanical sub-systems. It also shows the benefits of monitored compensation in contrast with the less efficacious and more man-power intensive trimming methods of the past. The case of drift compensation illustrates that this can only be done with respect to knowledge of the nature of the desired signal. Particularly notable, however, is the effectiveness of the removal of the unwanted time response by means of a tailored filter. The delicacy of this operation contrasts with the crudity of the traditional massive mechanical or electrical damping. The fundamental non-linearity of the weighing process is an area with which traditional techniques simply cannot cope, and the success of the adaptive filtering approach is an outstanding testament to the superiority of intelligent sensor systems.

One important advantage of the approach adopted here is that it is model-free (§ 3.2). True, we have in this case assumed a second order system, but that is not an essential restriction for reasons other than processing time. The optimization process is very efficient for the situation of figure 6.20, where batches of sensors come of a production line, as there is unlikely to be a great difference between adjacent devices, so we always start the process with a very good guess. The internal filter is then tailored to each individual primary sensor, with no preconceptions about its response. The adaptive filtering approach to noise suppression is also an important example. It show how noise has to be dealt with on a basis of knowledge of the nature of the desire signal (§ 3.2) and the same applies to drift correction. The assumption here is that the desired signal is a plateau; yet the method is still an improvement in dynamic weighing, such as the filling of cartons, though being non-linear it is difficult to analyse and is best treated by simulation.

If the authors were required to nominate one case to exemplify the advantages of intelligent sensor methods the load cell would be it.

## 8.3 AN OPTICALLY COUPLED PRESSURE SENSOR

### 8.3.0 Introduction

Fibre optic coupling offers a number of important advantages. First, it is inherently safe, in that the energy densities required for practical devices are well below the threshold required to ignite atmospheres containing explosive gaseous or particulate components. Second, it is immune from electromagnetic interference, as voltages and currents cannot be induced by stray fields. Third, it is robust and relatively immune to destructive forces, such as the extreme case of a large radiation flux in a nuclear explosion. These are formidable gains, but as in all engineering they have to be traded off against other complications. In this particular exercise we address two problems. How can we provide for both excitation of the sensor and detection of the response without duplicating fibres? How can we provide the vital monitoring line of figure 6.1 to implement monitored compensation for temperature cross-sensitivity?

### 8.3.1 Mechanisms

A primary sensor mechanism for pressure is the diaphragm (§ 3.3.1.1). This is one of our conceptual blocks (§ 3.1.1) that converts the variable pair of pressure and flow to the pair of stress and strain. Our standby element for the next stage of conversion to the voltage and current pair is the strain gauge (§ 3.3.1.1), and indeed many commercial pressure sensors take this route. In the present circumstance, however, we choose to eschew the electrical system and go for a conversion directly via the electromagnetic system of light, both for excitation and detection.

An element that converts stress and strain to another variable pair, tension and deflexion, is the vibrating string or beam. The stringed musical instruments make us familiar with the idea that we can excite vibrations at various harmonic frequencies and tune them via tension. An encastré beam is one of the simpler structures that can be fabricated by the microengineering of silicon (§ 5.2.3). By one of a number of etching techniques it is possible to etch a shallow pit into one side of a silicon diaphragm, but leave behind a beam that bridges it. When the diaphragm is subjected to a pressure differential the beam is stretched or compressed and its natural resonant frequency is altered in turn.

The next problem is how to excite the beam into a resonant motion at one of its natural frequencies. There are two possible effects of light that can be utilized, an electronic effect and a thermal one. If the silicon is covered with a thin layer of metal the thermal effect, aided by the bi-metallic strip phenomenon, can produce a usable action. A pulse of light shone on the structure produces a stress and strain the propagate along the beam as an acoustic wave. If pulses are repeated at a rate that corresponds to one of the natural frequencies of the beam resonant oscillation occurs.

If we are to use a single fibre for both excitation and detection it is necessary to time-multiplex these activities, and the way to do this is to use tone bursts to excite the resonance separated by spaces during which the response is measured. The beam is effectively a second order system (§ 2.2.6) and in this application it is convenient to rewrite the expression for the poles in terms of the resonant frequency $\omega_o$ and the $Q$ of the system. Provided $Q^2 \gg 1$ we can say that the poles occur at

$$s = -\frac{\omega_o}{2Q} \pm j\,\omega_o \ . \tag{8.1}$$

Thus the natural response is a sine wave of angular frequency $\omega_o$ with an exponential envelope of time constant $2Q/\omega_o$. The number of cycles per time constant is therefore $Q/\pi$. This is a critical number as it predetermines how many cycles are available for the detection process. Fortunately it is reasonably easy to fabricate a simple beam in silicon with a $Q$ of 100, so of the order of tens of cycles are available before the amplitude decays to a half. Thus if we turn the excitation on and off at a frequency of the order of one tenth of the resonant frequency, we can apply a detection process in the gaps, and to combat noise integrate the process over any number of gaps (Vincent 1993).

### 8.3.2 Monitored compensation

As we have observed (§ 3.2) monitored compensation is an essential requirement for an intelligent sensor, and a necessary part of the structure is the monitoring line of figure 6.1. How do we realize this within our optical system? The encastré beam is coupled to the variables of pressure (via strain) and temperature, so all we have to do is provide a beam with the former coupling removed. A cantilever beam fulfils this purpose. Thus the structure we require for the basic sensor is a silicon diaphragm in the centre of which is an etched pit across which are two beams, one encastré and the other cantilever. Such a structure is illustrated in the photograph of figure 8.8.

### 8.3.3 Information processing

This is an application where separation of the signal from noise is a very important aspect, but is also an example of a stimulus-response sensor (§ 6.5), and the knowledge that there is a correlation between the input and output signals provides the clue as to how to deal with the problem. In this system there is a number of noise sources, electrical, optical and acoustic, so the information processing stage is largely concerned with the recovery of the signal from noise. The encastré beam shown in figure 8.8 resonates at about 100 kHz with a $Q$ of about 100, so we can deduce from equation 8.1 above that, if we apply an optical signal of this frequency gated on and off at about ten kHz with unity mark-space ratio, the oscillation will be sufficiently established during the spaces for it to be detected. The knowledge that the response is coherent with the original signal, while the combined noise signal is uncorrelated, makes a number of powerful signal recovery processes possible (§ 4.4.5). In this case a variation of the phase sensitive detector was employed. the circuit being gated so that it is only operative during the spaces. Space precludes any detailed discussion of the optical and electronic systems, but these are described elsewhere (Vincent 1993).

### 8.3.4 Discussion

This particular study has underlined a number of important fundamental principles. It shows, for example, why it was necessary to include excitation control in our list

230  INTELLIGENT SENSOR SYSTEMS

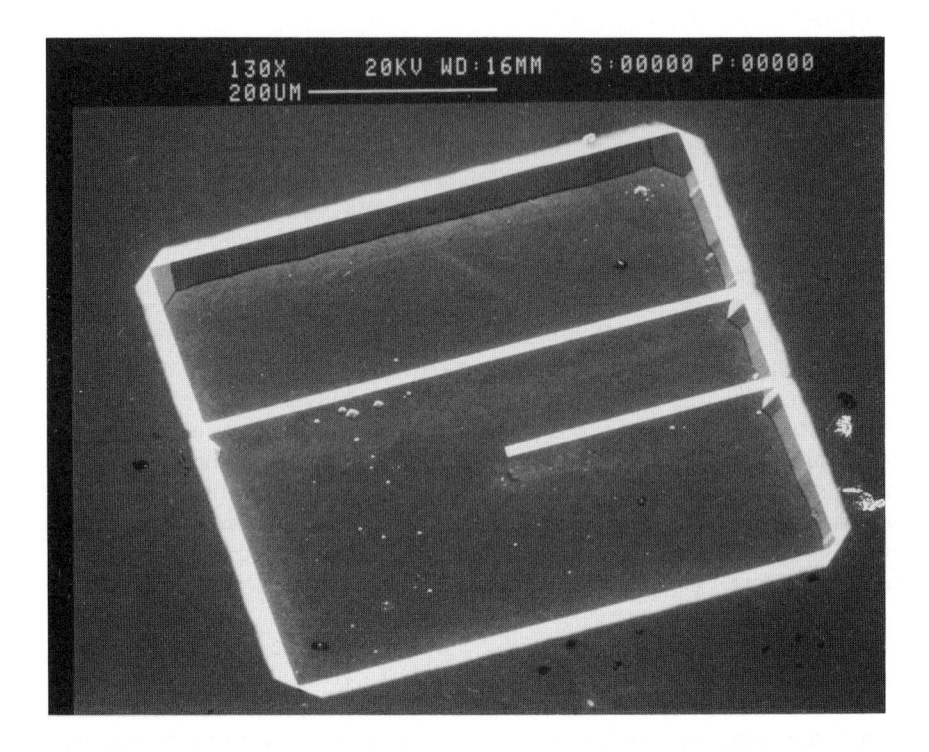

**Figure 8.8** Photograph of the dual beam structure etched into the surface of the pressure diaphragm.

of elements of intelligent sensors (§ 6.1) and justifies the remark that the possibilities of this aspect are largely unexplored. It is also unique in this context for the fact that the primary sensor and the intelligence unit are physically separated, perhaps by a large distance. Although we have tended to define an intelligent sensor as one with the processing contained in its own housing, none of the basic principles laid down here is affected by such separation.

The optical fibre offers enormous advantages where safety critical aspects are important. There are three basic ways in which it can be exploited - the intrinsic fibre sensor, the optically powered electronic system (§ 4.1.3) and the case above in which the stimulus-response mechanism is wholly optical. Space considerations preclude more detailed discussion here, but there is a copious literature (e.g. Gambling 1991).

## 8.4 THE ELECTRONIC NOSE

### 8.4.0 Introduction

One of the most challenging applications of sensors has been the detection and classification of odours. In the preface to this book we referred to the animal nose as an example of the remarkable effectiveness of evolution in producing superb sensors. Man is now attempting in a modest way to emulate these achievements.

### 8.4.1 Mechanisms

As we have seen in our earlier discussion (§ 3.3.3) there has been a great deal of research and development activity in the area of chemical sensing. It is possible to make quite simple devices with a very high sensitivity to incoming gases and vapours. Unfortunately this sensitivity is not unique, so much so that the problem of cross-sensitivity is dominant and the individual devices are in themselves virtually useless. In this field the problem is referred to as one of specificity. Two concepts make progress possible. The first is the idea of the sensor array and the second is our predominant theme of digital information processing. In our preliminary discussion of sensor array processing (§ 6.3.4) we drew attention to the magnitude of the problems involved, virtually every one in the book. This is a highly interdisciplinary subject, much of which is outside the remit of this text, and indeed the competence of the authors. We are grateful in preparing this short summary to have had access to material supplied by workers at the universities of Southampton and Warwick.

The basic device is the chemiresistor (§ 3.3.3.1). Its complications include a poor selectivity and a strong cross-sensitivity to temperature. The latter, however, can be turned to advantage; for if this parameter is controlled it give a variability of chemical response which reinforces the variability obtained by variation of materials. The sensor array, then, is a single substrate with a number of different chemically sensitive materials operated at a number of temperatures (figure 3.25)

The most mature technology for primary gas sensors is based on metal oxides, arising from the discovery by Bratten and Bardeen (1953) that adsorption of a gas on the surface of a semiconductor can produce a significant change in electrical resistance. A variety of materials have been used (e.g. $ZnO_2$, $TiO_2$, $WO_3$ ) but by far the most successful has been sintered Tin Oxide doped by a precious metal (Gardner 1991). Subsequently organic semiconductors, such as the phthalocyanines, have become important.

### 8.4.1 Structural compensation

The most important aspect of structural compensation for gas sensor arrays is the physical size. Typically, interdigitated electrode structures are fabricated on an alumina substrate by thin-film (§ 5.3), or thick-film (§ 5.4) techniques. In a typical structure (figure 3.25), each site has an individual heater and is thermally isolated from its neighbours by slots in the substrate. Compactness is necessary to minimize the effects of local variations in chemical concentration.

### 8.4.2 Monitored compensation

As stated above temperature is an important parameter, but the approach in this case is to control it rather than monitor it. Nevertheless, in order to control we need to measure, and the resistance of the heater element provides a convenient intermediary for this purpose. The very concept of a sensor array itself is a variation of the principle of monitored compensation, and it is only conceivable in the concept of an intelligent sensor system.

### 8.4.3 Information processing

It is impossible to encapsulate the methods of information processing used in electronic nose applications with the brevity that is necessary here. The scale of the problem can be conceived by considering the mammalian nose. This can identify as little as one part in $10^{12}$ concentrations of odorous substances by way of 50 million olfactory receptor cells synaptically linked into several thousand glomeruli which in turn feed into the mitral cells in the olfactory bulb in the brain (Gardner and Bartlett 1992). The practical difficulties posed by the inadequacies of our primary sensors (§ 6.3.4) pose a problem of daunting complexity and it is a mark of the ingenuity of workers in the field that such successful results have been obtained.

The basic problem is one of pattern recognition (PARC). We start out with a vector of sensor responses and have to implement a block process (§ 6.2.2.3) like that of equation (6.2), except that we do not have the benefits of linearity. It is similar to our problem of waveform recognition and classification (§ 6.3.4.7) and, as there, we can devise various metrics to separate out the target chemicals.

There are two basic schemes of pattern recognition employed in electronic noses, the *supervised learning scheme* and the *unsupervised identification scheme* (Gardner and Bartlett 1992). The architectures for these are shown in figure 8.9. It will be seen that the process can be divided into two transformations. The first is from a pattern space to a feature space and the second is from the latter to a classification space. There is a variety of methods for performing these transformations, and it is largely the first that characterizes a particular scheme. Let us

## PHYSICAL REALIZATIONS 233

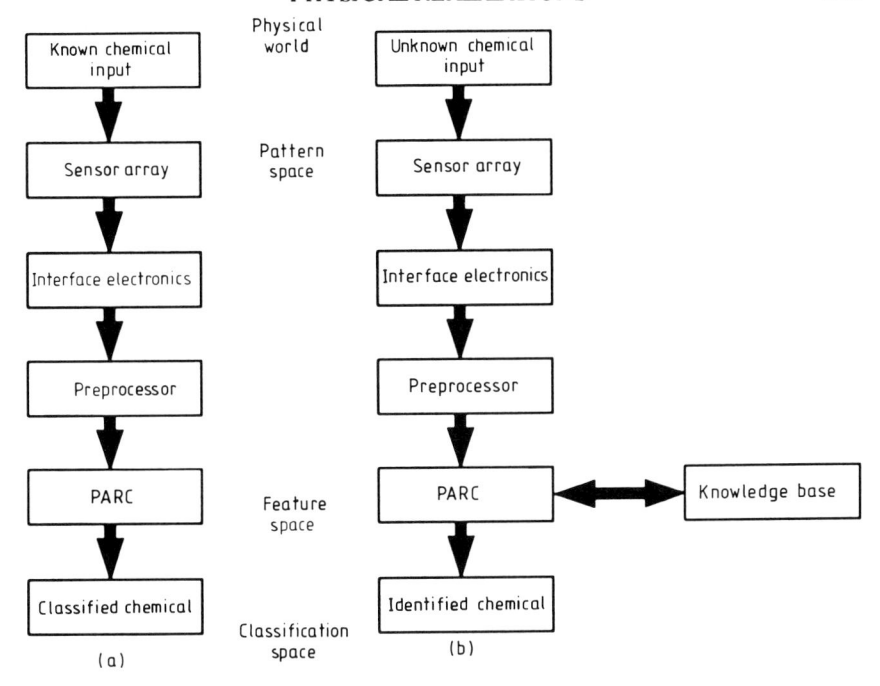

**Figure 8.9** Generalized pattern recognition architectures for the electronic nose, (a) a supervised learning scheme and (b) an unsupervised identification scheme.

examine just one of these as an example, noting that there are many other variations in the literature.

Principal Component Analysis (PCA) is a supervised pattern recognition technique. Internally it is a linear process immersed in a non-linear environment. The process is similar to that of finding an orthogonal basis in a vector space (Gardner and Bartlett 1992). In general the input variables are normalized to zero mean and unit variance.

If vector $x$ represents the normalized array of input responses from the sensor we form a new orthogonal basis $y$ by means of a linear transform on $x$ in a way analogous to our block process (§ 6.2.2.3, equation (6.2)). There are classical mathematical methods (Mirsky 1955) of finding the coefficients $b$ in the transform.

$$y = b^T x \tag{8.2}$$

Consider a simple geometric analogy. Imagine a number of objects in three-dimensional space differentiated in some way, say by colour. If we look at these objects from every possible angle there is likely to be one angle from which the groups of objects are maximally discriminated. If we now take a colour photograph from that angle, effectively projecting the space onto a plane, we illustrate that discrimination. The plane can be defined by two orthogonal axes, which are our principal components. Surprisingly enough two principal components are often

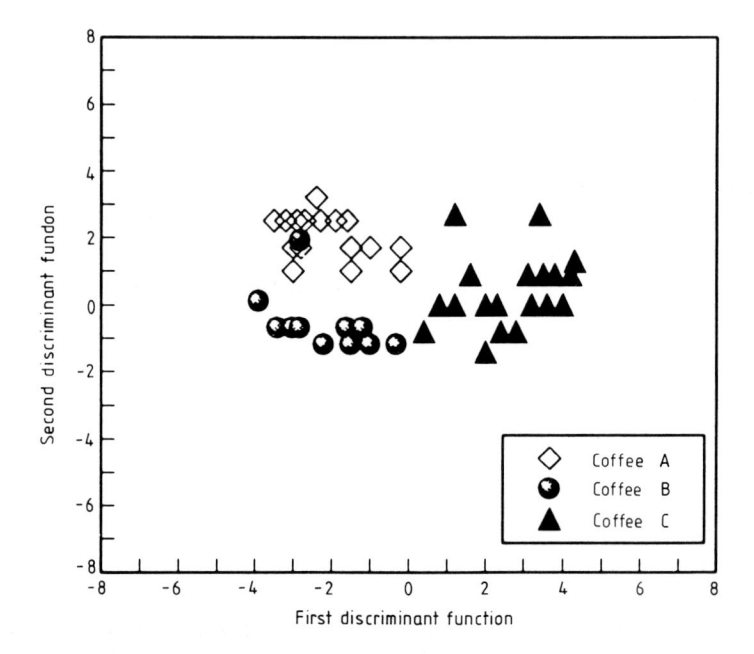

**Figure 8.10** Results of discriminant function analysis on three commercial coffee odours (after Gardner and Bartlett 1992).

sufficient to show this discrimination even though the original number of sensors is considerably greater than two. The idea of supervised learning is that we apply a series of standard chemicals to the system to find the principal components, and work on the hypothesis that the separation will also be maximally effective for other samples.

In order for such a linear process to be effective in the non-linear environment it is normally necessary for there to be a preprocessor stage, which may not be linear. Figure 8.10 shows one example of how just two discrminants are able largely to differentiate three brands of coffee with differing blend or roasting level. Here practice varies widely and various transformations including logarithmic are performed on the incoming information of resistance or conductance.

We have, all too superficially, described just one of many forms of pattern recognition as an exemplar which happens to relate to earlier comments in this book. Some of the techniques are based on less traditional computational methods that we have not been able to cover, such as neural nets and fuzzy logic. Neural nets seem to be particular promising in this area of application (Gardner *et al* 1990). Considering the disadvantages the electronic nose faces in comparison with the mammalian variety there have been some notable successes. It has proved possible, for example, to discriminate between various varieties of coffee or beer and automatically identify faulty batches.

### 8.4.4 Discussion

This last case study is again quite different from the three predecessors, illustrating the variety of implementations possible for intelligent sensor techniques, yet also underlining some of the common features. The basic forms of sensor compensation (§ 3.2) are still present, though in very different guises. The single primary gas sensor is so non-specific as to be virtually useless; yet, by combining the physical array of sensors with the idea of array processing, it can be turned into a powerful tool of measurement. The electronic nose has developed out of unpromising material into a significant advance in instrumentation.

## 8.5 ASICs

### 8.5.0 Introduction

It is odd that when, inevitably, the discussion of the next stage of intelligent sensor development turned to Application Specific Integrated Circuit (ASIC) technology, the same cost arguments used to decry the original concept were trotted out. Again the opponents were guilty of looking only at the present cost situation and were ignoring the trends. Those involved in research and advanced development, however, have to be guided by the trends and set their sights a decade into the future. These trends were quite clear and established almost from the first appearance of transistorized electronics, and particularly of integrated circuits (Brignell 1993).

The cost of a logic transistor has been halving every one and a half to two years, while the density of transistors has been doubling every three years. Furthermore, the size of viable chips has also been steadily increasing. As a result any device which is state of the art in power at one time ( say a million transistor processor ) is a decade later merely a small sub-system and heading for one hundredth of the price. Thus we head to the point where the silicon comes free; though nothing else does, including packaging, interconnection, testing and, above all, design.

Perhaps the only nearly similar experience has been the ballpoint pen. When first launched Biro's invention was the plaything of the wealthy. Now it is given away with other products, such as sensor conferences. This is not to say that it is any less useful, merely that its production costs have dwindled almost to insignificance.

In like manner we have to prepare for a world in which **the silicon comes virtually free**. The result is a change in the relationship between the fixed and variable costs of production, as illustrated diagrammatically in figure 8.11 (Brignell 1993). Silicon technology is then no longer constrained to serve large scale production, but becomes viable for medium and small runs. There is one constraint,

*236* *INTELLIGENT SENSOR SYSTEMS*

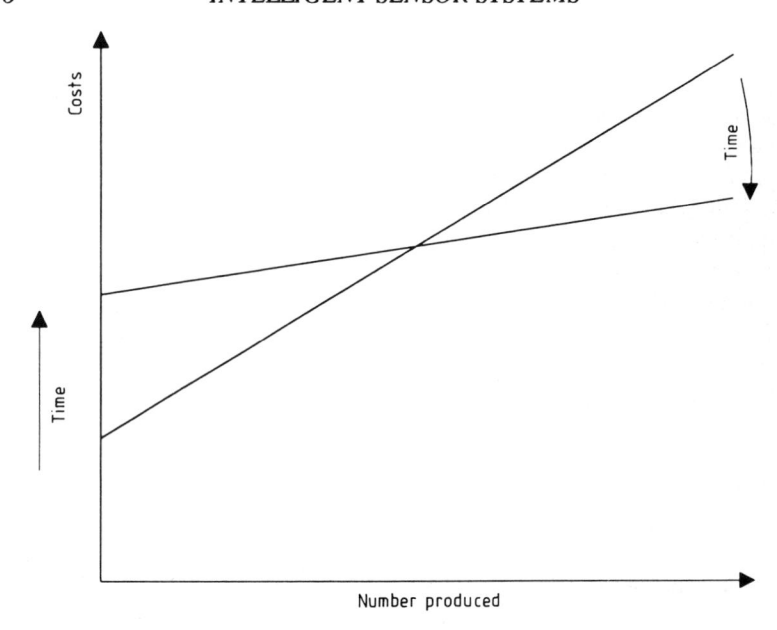

**Figure 8.11** A crude illustration of how the costs of production are changing with time.

**Figure 8.12** The roving slave processor, an early attempt to take sensors out into the field (Brignell *et al* 1976, Brignell and Buffam 1977).

*PHYSICAL REALIZATIONS* 237

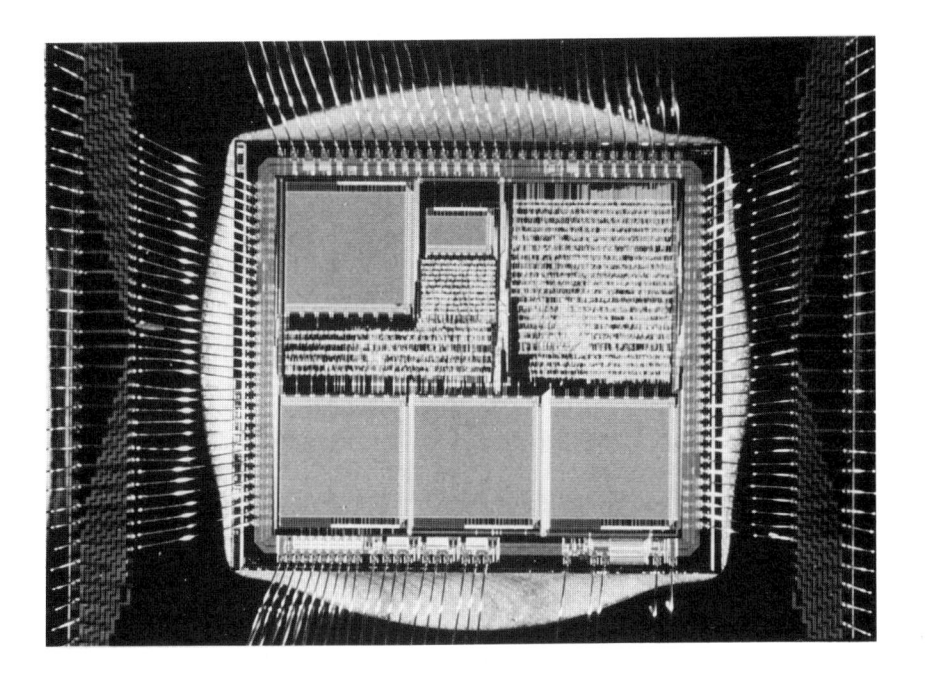

**Figure 8.13** An ASIC for intelligent sensor applications; incorporating an embedded microprocessor, memory and analogue circuits.

however, and this statement is only true **if we can contain the fixed costs**, especially the all important design costs.

The rate of progress of the technology is nicely illustrated by photographs of realizations of intelligent sensor sub-systems associated with the authors' research taken two decades apart. Figure 8.12 shows one of the early attempts to take digitally supported sensors out into the field. This system from the early 1970s (the roving slave processor (Brignell *et al* 1976) ) was based on one of the first sixteen bit microprocessors to become available (Ferranti F100L). With support systems, including power supplies, data conversion and non-volatile magnetic bubble memory, however, the physical realization comprised 19 inch racks on a frame with castors.

Twenty years later a system with equivalent power can be housed in a single chip that can be powered by a modest battery. Figure 8.13 shows such a chip that was fabricated while this text was in the final stages of preparation. It contains data conversion and an embedded microprocessor, in fact the complete electronic sub-system to realize and intelligent pressure sensor. The chip was designed and fabricated by Lucas and ES2 as part of a LINK programme supported by DTI and SERC in which the authors' laboratory were partners. This illustration of the dramatic rate of change in the technology serves to remind researchers that it is not good enough to pitch their efforts at the current state of the art. Now, as in the early days of intelligent sensors, the technology that appears to be outrageously expensive today becomes the accepted norm within a few years. At the beginning of the 1990s ASIC technology was regarded as expensive and only suited to relatively large production runs, but this situation has been changing rapidly.

### 8.5.1 A strategy for intelligent sensor ASICs

The early history of computer software gives us a clue to the approach required in our area of technology. The first computers were programmed at a very low level, machine code, which was difficult to develop and modify. When a new computer appeared we threw all the previous programs away and started all over again. This 're-inventing the wheel' was an inordinately wasteful process, which was gradually eliminated by the move to higher levels of language, through assemblers to the more powerful human-oriented forms such as FORTRAN, ALGOL and PASCAL. With these came modularity, aids for design and testing. By the time microprocessors appeared there was a tendency to repeat this process, which was rapidly quelled by the needs of for greater productivity from a restricted pool of skilled people. So other concepts appeared, such as emulation and simulation, along with portability and rigour of language.

In our strategy for ASICs in intelligent sensors we must seek to side-step these early stages and proceed to the advanced concepts. the following are some of the components that lead to such a strategy.

#### 8.5.1.1. A design philosophy
There is a tendency for inchoate technologies also to be chaotic. If this can be avoided a great deal of human effort can be saved. The intelligent sensor comprises a number of subsystems linked together (figure 6.1). This can obviously be achieved in a number of different ways. One important aspect of a design philosophy is the creation of a standard framework on which the sub-systems can be mounted. An example of such a framework is to shown in figure 8.14, in which figure 6.1 is modified by the addition of a control processor. This can be a simple sub-system which does not have to of high performance as it serves to control the settings of the other sub-systems (gain, signal source selection, offset, communication control etc.). Its presence, however, does provide a skeleton upon which the flesh of the design can be imposed, as can be seen from the structure of dotted lines in the figure.

#### 8.5.1.2 Stop re-inventing the wheel
As we have have observed, in the early days of computing a great deal of human effort was wasted by a process of throwing away everything that had been done in order to start all over again. It is clearly possible to start every intelligent sensor ASIC from scratch and design a complete system to solve a particular problem. this is, however, a wasteful process. To avoid this we have to adopt a number of approaches. First, we conceive the system as a set of subsystems according to a

*PHYSICAL REALIZATIONS* 239

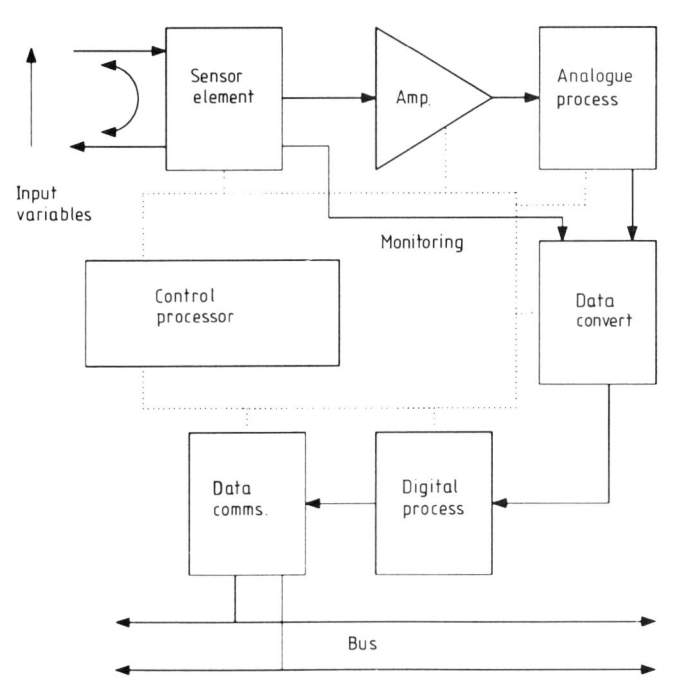

**Figure 8.14** A revised version of figure 6.1, in which a control processor becomes the skeleton upon which an ASIC framework can be built.

standard scheme such as that illustrated in figure 8.14. Second, having designed a subsystem to fulfil a particular requirement we save it as a library element for future use. This implies a third requirement, that we have to create standards by which such subsystems are linked together. This is particularly important in those paths concerned with signal flow. Here techniques such as the cyclic buffer (§ 6.2.2.2) help by providing *elasticity*. Obviously there are various levels of skill required in producing subsystems. and this leads to the next aspect of the philosophy.

#### 8.5.1.3 De-skilling the design process

There is no doubt that there is a basic requirement for some highly skilled design activity, and there will always be a need for a number of people to be involved at the lowest (transistor) level of design. In general, however, it would be too much to expect instrumentation engineers to add to their skills the demanding requirements of integrated circuit design. Likewise, experienced circuit designers would have a long way to go before they mastered all the requirements of instrumentation systems engineering. Instrumentation engineers are used to accepting packaged chips and treating them as black boxes fully specified by their published characteristics, though we have all experienced hazards in doing so.

The idea of accepting sub-systems from a library as fully specified black boxes is little (but not no) different from what we are used to. A necessary pre-requisite, however, is that such sub-systems are designed by skilled multi-disciplinary teams

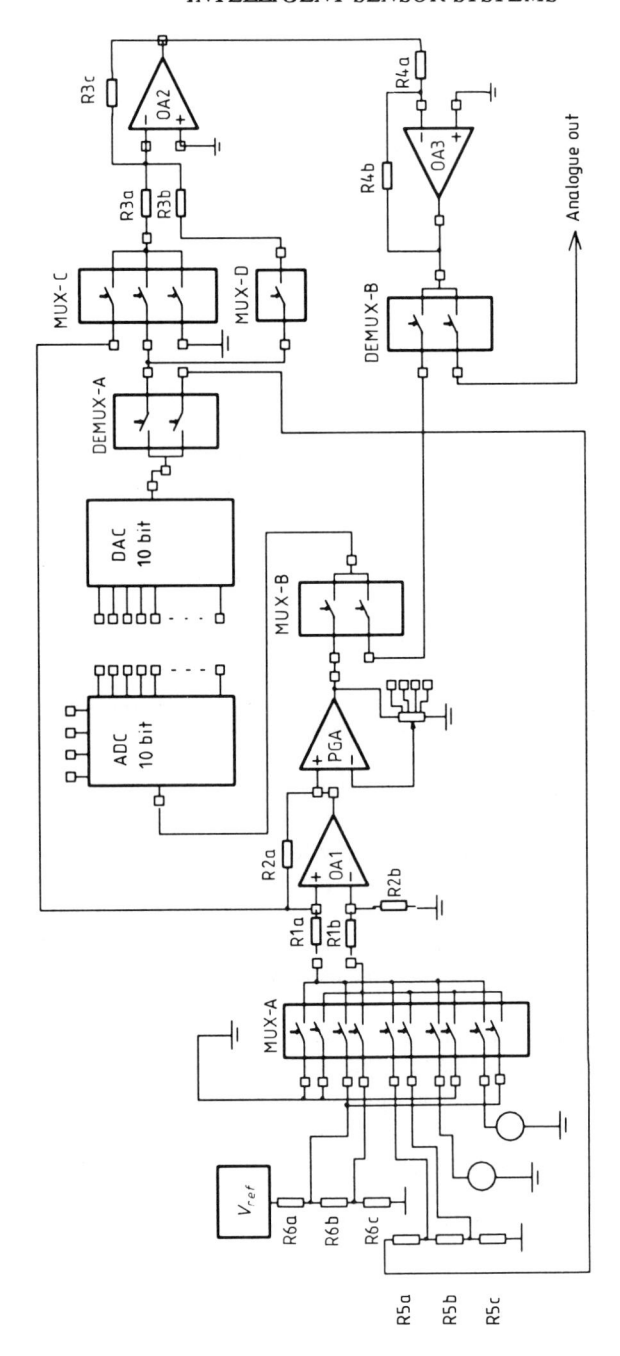

**Figure 8.15** Layout of a test chip to try out various strategies for self-test and self-calibration (after P Yick and J Kruger).

who appreciate equally the instrumentation requirements and the circuit capabilities. Basic circuit technologies optimized for compact, high-speed digital circuits do not lend themselves easily to the demands of analogue instrumentation sub-systems ( linearity, low noise etc.). This, of course was a major part of the case for thick-film hybrid techniques to combine technologies optimized for their own particular function. New technologies are emerging which promise to overcome these problems, but in the mean time special design techniques have to be employed ( such as the use of switched capacitors to overcome problems of offsets etc). For the non-specialist to be able to put together ASICs he requires a number of facilities embedded in a coherent design philosophy, as we have observed above. One existing problem is that the simulation of analogue circuits, and particularly mixed analogue-digital circuits is not as advance as the digital case, where simulation results can be a guarantee of correct operation of the real device. Testing of such mixed circuits is a demanding task, from which the average instrumentation engineer. Figure 8.15, for example shows a fully designed version of the rather naive system of figure 6.4. This design was used in the authors' laboratory to try out various self-test and self-calibration procedures using real ASIC sub-systems with all their imperfections, and as such it had a large number of controlled switches.

The hardware and software aspects of testing a system like that of figure 8.15 can be very demanding. Not because they are difficult but rather because they are complex and require a great deal of care in ensuring that all relevant conditions have been applied. The sheer number of connections can create there own problems, and one of the main motivations towards a single chip solution is the desire to reduce this number; for connections not only cost money in manufacture and test, they also reduce reduce reliability.

A word on the approach to testing such a chip would not be out of place. There are, of course, all sorts of elaborate special functional testers available, and these are of great value. But ultimately we would wish to test the sub-systems in a context similar to the ones in which they would be used. This not only includes the specification of the hardware drivers but also the software ones. Thus, simultaneously with the testing of this chip software drivers were developed in the C language for portability. They were originally implemented in a PC, so the obvious from of test-bed for the device is a specially designed card suitable for inserting in the PC itself. The signal levels were perforce those that would be used within a chip, and the driver sub-programmes were suitable for transferring to an embedded processor in a later realization. Figure 8.16 shows the chip mounted in a PC card.

But why, the reader may ask, do we include all this detail in a section entitled 'de-skilling the design process'? The reason is to illustrate the procedures that have to be gone through, but from which the end user should be protected. Not only the hardware modules, but also the software ones, should be fully tested, user-friendly and compatible. Thus a self calibrations subsystem should be available as a library module that can be incorporated in future designs, and also a software module which conceals any complexities of operation, while offering a simple interface to the applications software or firmware in the embedded processor.

**Figure 8.16** The test chip of figure 8.14 mounted in a PC card.

### 8.5.2 Discussion

As elsewhere in this text we have only been able to give the briefest account of this important area, and to do otherwise would require a whole new book. What we have tried to do, however, is to try to place it in a context as a natural development from what has gone before, and to emphasize that the fundamental principles still apply. Elementary ideas, like the sampling theorem, retain their importance, but can easily be forgotten when one is trying to cope with a complex design procedure. This, above all, is a justification for the requirement to de-skill this process, as emphasized above. The powerful technologies now available to us can bring confusion instead of advance unless we continually strive for conceptual simplicity by building a hierarchy of self contained and tested sub-systems. For a system to be correct it is not a *sufficient* condition that its sub-systems be correct, but it is certainly a *necessary* one.

# 9
# Conclusion

In these pages we have progressed, in a fairly continuous fashion, from a few very broad, simple principles to some advanced implementations. Elementary questions in chapter one (What is measurement? What is information?) have led via considerations of systems theory, physical mechanisms, enabling technologies, digital and analogue techniques, to an account of practical devices which dramatically enhance our ability to measure what is happening in the real world.

That most basic principle of mathematics, the fundamental theorem of algebra (§ 2.1.5), held the seed that would grow into a solution of one of the most intractable problems of sensing, eliminating the resonant response of a mechanical sensor; but that growth could only occur when fed and watered by the combination of rigorous systems theory with powerful cheap and compact electronic technologies. Comparing the imposition of the traditional massive mechanical damping with the modern adaptive filtering techniques is like comparing the butchery of surgeons in the wooden ships with modern microsurgery; yet, like all electronic advances, it has happened within a generation.

The purpose of this text has been to emphasize that intelligent sensing is a discipline in its own right, or perhaps more accurately a multi-discipline. The basic principles become even more important when we have powerful tools at our disposal, for it is so easy to misuse those tools in ignorance of the fundamentals. The competent digital instrumentation engineer never forgets that he has grossly corrupted the input signal by sampling and quantization, yet much digital design is still carried out via naive analogues of the continuous case. It is not difficult to imagine cases where the consequences of such artlessness could be literally disastrous.

Another reason for commending the study of this subject is that it is a very good didactic tool in the development of young engineers. The wide span of inter-related subject matter, the repeated appearance of important engineering trade-offs and the harnessing of a variety of technologies to achieve stated objectives are all excellent preparation for an engineering career. Students have often remarked when they come to this subject that it makes them realize what all that previous theory was for.

We began chapter one with the remark that the existence of measurement was a necessary condition for the development of civilization. As civilization becomes more complex, so more complex measurement is required. We can only confidently control damage done to the environment if we can measure that damage, be it monitoring the gases in an exhaust pipe or the hole in the ozone layer. Increasingly the economic well-being of a company or a nation depends on the quality of their products. Determination of that quality is a measurement process. The more complex society becomes the greater the potential for disaster. The key to fending off disaster is information, and the key to information is measurement. Man is fundamentally a tool maker and he can use his tools for good or ill. Intelligent sensor systems are powerful new tools which promise great good for mankind if wisely used.

# References

*Manufacturing Automation Protocol Reference Specification (Society of Manufacturing Engineers) (Dearborn 1988)*
*Reference Model of Open Systems Interconnection ISO/TC97/SC/6 D/S 7498 1983 (International Standards Organisation)*
Atkinson J K 1991 *Proc. Advances in Analogue and Digital Interfacing of Transducers (Southampton 1991)* (Southampton: USITT)
Barbe B F 1975 *Proc. IEEE* **63** 38-67
Bart S F, Tavrow L S, Mehregeny M and Lang T H 1990 *Sens. Actuators* **A 21-23** 193-7
Baustra S B, Legtenberg R, Tilmans H A L and Elvenspoek M 1990 *Sens. Actuators* **A 21-23** 332-5
Bentley J P 1988 *Principles of Measurement Systems* 2nd edn (Harlow: Longman)
Bergveld P 1970 *IEEE Trans. Biomed. Eng.* **19** 70-1
Bergveld P 1985 *Sens. Actuators* **8** 109-27
Berry R W, Hall P M and Harris M T 1968 *Thin Film Technology* (New Jersey: Van Nostrand)
Birch R D, Payne D N and Varnham M P 1982 *Electron. Lett.* **18** 1036-8
Blasquez P, Pons P and Boukabache A 1989 *Sens. Actuators* **17** 387-403
Bradley D A, Dawson D, Burd N C and Loader A J 1991 *Mechatronics – Electronics in Products and Processes* (London: Chapman and Hall)
Brignell J E 1984 *J. Phys. E: Sci. Instrum.* **17** 759-765
Brignell J E 1985 *J. Phys. E: Sci. Instrum.* **18** 559-565
Brignell J E 1987a *J. Phys. E: Sci. Instrum.* **10** 1097-1102
Brignell J E 1987b *Sens. Actuators* **10** 249-261
Brignell J E 1989 *Sensors a Comprehensive Survey:* Volume 1 *Fundamentals and General Aspects* ed W Gopel, J Hesse and JN Zemel (Germany: VCH)
Brignell J E 1991 *Sens. Actuators* **A 25-27** 29-25
Brignell J E 1993 *Sens. Actuators* **A 37-38** 6-8
Brignell J E and Buffam C J 1977 *Proc. IERE Conf (No 38) on Programmable Instruments* 7-16
Brignell J E, Comley R A and Young R *1976 Microprocessors* **1** 79-84
Brignell J E and Dorey A P 1983 *J.Phys. E: Sci. Instrum.* **16** 952-958
Brignell J E and Rhodes G 1975 *Laboratory Online Computing* (London: Intertext)

Brignell J E, White N M and Cranny A W J 1988 *IEE Proc. I* **135** No.4 77-84
Brignell J E and Young R 1979 *J. Phys. E: Sci. Instrum.* **12** 455-463
Buffam C J 1976 *PhD Thesis* City University, London
Burley 1974 *Studies in Optimisation* (London:Intertext)
Buttle A, Constantinides AG and Brignell J E 1968 *Electron. Lett.* **4** 252-253
Catteneo A, Dell 'Acqua R, Forlani F and Pirozzi L 1980 *Society of Automotive Engineers (Detroit)* 49-54
Chowdhry B S 1989 *PhD Thesis* University of Southampton
Chowdhry B S, Shahi S and Brignell J E *1988 J. Phys. E: Sci. Instrum.* **21** 259-263
Christel L, Petersen K, Barth P, Pourahmadi F, Mallon J and Bryzek J 1990 *Sens. Actuators* **A 21** 84-8
Collet C V and Hope A D 1983 *Engineering Measurements* 2nd edn (Harlow: Longman)
Collet M G 1986 *Sens. Actuators* **10** 287-302
Comley R A and Brignell 1981 J E *J. Phys. E: Sci. Instrum.* **14** 963-967
Conklin G (ed) 1962 *Great Science Fiction by Scientists* (London: Collier-Macmillan)
Constantinides A G 1968 *Electron. Lett.* **4** 115-116
Cooper AR and Brignell JE 1984 *J. Phys. E Sci. Instrum.* **17** 627-628
Cooper A R and Brignell J E 1985a *Sens. Actuators* **7** 189-98
Cooper A R and Brignell J E 1985b *J. IERE* **55** 7/8 263-7
Dalley J W and Riley W F 1978 *Experimental Stress Analysis* (New York: McGraw-Hill)
Doebelin O D 1990 *Measurement Systems Application and Design* (Singapore: McGraw-Hill)
Finkelstein L and Watts 1971 *Meas. Control* **4**
Gambling W A 1991 *Sens. Actuators* **A 25-27** 191-6
Gardner J W 1991 *Sens. Actuators* **4** 109-116
Gardner JW and Bartlett PN 1991 *Techniques and Mechanisms in Gas Sensing* ed PT Moseley, J Norris and DE Williams (Bristol: Adam Hilger)
Gardner J W and Bartlett P N (eds) 1992 *Sensors and Sensory Systems for an Electronic Nose* (Dordrecht: Kluwer Academic Publishers)
Gold B and Rader C M 1969 *Digital Processing of Signals* (New York: McGraw-Hill)
Goodyear C C 1971 *Signals and Information* (London: Butterworths)
Gumbel E J 1958 *Statistics of Extremes* (New York: Columbia University Press)
Holmes P J and Loasby R G 1976 *Handbook of Thick-Film Technology* (Glasgow: Electrochemical Publications)
Horowitz P and Hill W 1989 *The Art of Electronics* (New York: Cambridge)
Holford K M, White N M and Bakopoulos C 1990 *Proc. I. Mech. E Conf. Mechatronics* 47-59
Johnson L W and Riess R D 1977 *Numerical Analysis* (Reading, MA: Addison-Wesley)

# REFERENCES

Jones B E 1980 *Instrumentation Measurement and Feedback* (New Delhi: McGraw-Hill)
Kissinger P T and Heinman W R 1984 *Laboratory Techniques in Electro-Analytical Chemistry* (New York: Dekker)
Koo K P and Siegel G H 1982 *IEEE J. Quantum Electron.* **18** 670
LaFara R L 1973 *Computer Methods for Science and Engineering* (Princeton, NJ: Hayden)
Laming R I, Payne D N and Li L 1988 *Conf. Optical Fibre Sensor (New Orleans 1988)* **2** 123-8
Leaver, L D and Chapman B N 1971 *Thin Films (The Wykeham Science Series)* ( London: Wykeham)
Loxton R and Pope P (eds) 1986 *Instrumentation: A Reader* (Milton Keynes: Open University)
Mallon J R, Pourahmadi F, Petersen K, Barth B, Vermeulen T and Bryzek J 1990 *Sens. Actuators* **A 21-23** 89-95
Mayhan R J 1983 *Discrete Time and Continuous Time Systems* (Reading, MA: Addison-Wesley)
Middelhoek S and Audet S A 1989 *Silicon Sensors* (London: Academic)
Mirsky L 1955 *An Introduction to Linear Agebra* (Oxford:Clarendon)
Morris A S 1988 *Principles of Measurement and Instrumentation* (London: Prentice-Hall)
Moseley P T and Tofield B C 1987 (eds) *Solid State Gas Sensors* (Bristol: Adam Hilger)
Moseley P T, Norris J O W and Williams D E 1991 *Techniques and Mechanisms in Gas Sensing* (Bristol: Adam Hilger)
Muller R S 1990 *Sens. Actuators* **A 21-23** 1-8
Nelder J A and Mead R 1965 *Comput. J.* **7** 308-313
Neubert H K P 1963 *Instrument Transducers* (Oxford: Oxford University Press)
Norton H N 1982 *Sensor and Analyzer Handbook* (New York: Prentice-Hall)
Ohba R 1992 (ed) *Intelligent Sensor Technology* (Chichester: John Wiley)
Oppenheim A V, Willsky A S and Young I T 1983 *Signals and Systems* (London:Prentice-Hall)
Papoulis A 1965 *Probability, Random Variables and Stochastic Processes* (New York:McGraw-Hill)
Papoulis A 1980 *Circuits and Systems* (New York: Holt, Reinhart and Winston )
Pickard R S, Wall P, Ubeid M, Ensell G and Leong K H 1990 *Sens. Actuators* **B1** 460-3
Poole S B, Payne D N and Fermann M E 1985 *Electron. Lett.* **21** 737-8
Prudenziati M, Morten B and Taroni A 1981 *Sens. Actuators* **2** 17-27
Prudenziati M and Morten B 1986 *Sens. Actuators* **10** 65-82
Prudenziati M and Morten B 1992 *Microelec.* **23** 133-41
Pucknell D A, Eshraghian K 1987 *Basic VLSI Design* (Englewood Cliffs, NJ: Prentice-Hall)
Rabiner R and Rader C M 1972 *Digital Signal Processing* (New York:IEEE Press)

Reynolds Q M and Norton M G 1985 *Proc. Test and Transducer* (Wembley 1985) **2** 31-44 (Tavistock: Trident)
Roberts G (ed) 1990 *Langmuir-Blodgett Films* (New York: Plenum)
Ross J N 1992 *Meas. Sci. Technol.* **3** 651-5
Schlabach T D, Rider D K 1963 *Printed and Integrated Circuitry* (New York: McGraw-Hill)
Sheingold D H (ed) 1980 *Transducer Interfacing Handbook: A Guide to Analog Signal Conditioning (Analog Devices Technical Handbooks)* (Norwood: Analog Devices)
Shepherd I E 1981 *Operational Amplifiers* (London: Longman)
Shi W J 1992 *PhD Thesis* University of Southampton
Shi W J and Brignell J E 1991 *Sens. Actuators* **A 25-27** 37-41
Shi W J, White N M and Brignell J E 1993 *Sens. Actuators* **A37-38** 280-5
Terry S C, Jerman J H and Angell J B 1979 *IEEE Trans. Electron Devices* **26** 1880-6
Van Putten A F P 1988 *Electronic Measurement Systems* (Hemel Hemstead: Prentice-Hall
Vincent D R 1993 *PhD Thesis* University of Southampton
Vincent P 1991*Proc. Advances in Analogue and Digital Interfacing of Transducers (Southampton 1991)* (Southampton: USITT)
Watson J 1989 *Analog and Switching Circuit Design* 2nd edn (Singapore: Wiley)
White N M and Cranny A W J 1987 *Hybrid Circuits* **12** 32-5
White N M 1988 *PhD Thesis* University of Southampton
White N M 1989 *Hybrid Circuits* **20** 23-7
White N M and Brignell J E 1991 *Sens. Actuators* **26** 1/3 313-9
White N M 1992 *Proc. ISHM Conf. Thick-Film Sensors (Southampton 1992)*
Wieder H H 1971 *Hall Generators and Magnetoresistors* (London: Pion)
Wilmshurst T H 1985 *Signal Recovery From Noise in Electronic Instrumentation* (Bristol: Adam Hilger)
Wobschall D 1979 *Circuit Design for Electronic Instrumentation : Analog and Digital Devices From Sensor to Display (*New York: McGraw-Hill)
Tarui Y (ed) 1981 *VLSI Technology-Fundamentals and Applications* (Tokyo: Springer-Verlag)

# Index

Accelerometer, 56, 196
    cantilever beam type, 57
    piezoelectric, 58
    reluctive, 59
Accuracy, 2 - 3
Actuator
    definition, 33
ADC, 2, 5 - 6, 97, 100, 102 - 110, 151, 226
    dual slope, 105 - 108
    parallel encoder (flash converter), 108
    ramp, 102
    successive approximation, 103 - 105
Addressability, 145
Admittance, 31
Alias, 26, 144, 161
Amplification, 143 - 144
Amplifier
    charge, 99 - 100
    high performance, 95 - 96
    instrumentation, 94 - 95, 146
    isolation, 96 - 97
    logarithmic, 97 - 98
    operational, 93 - 94
Analogue-to-digital converter, *(See* ADC)
Analysis
    qualitative, 71
    quantitative, 71
ASIC, 109, 217, 235, 238 - 242
Auto-calibration, 147
Autocorrelation, 18 - 20, 22, 200
Automatic breaking system (ABS), 194

Average, 27
    ensemble, 19 - 20
    moving, 27
    time, 19, 21 - 22
Averaging
    signal, 38 - 39, 113 - 114

Beam
    cantilever, 55, 59, 129, 229
    coupled, 56, 222
    encastré, 55, 228 - 229
    resonant, 228
Bit
    definition, 4
    stuffing, 208 - 209
Bounce, 189
Bridge
    capacitance, 91
    non-linear elements, 91
    strain gauge, 222
    Wheatstone, 63, 88 - 91, 198
Buffer
    cyclic, 155 - 156, 239
    overflow, 7
Building
    intelligent, 190
Bus, 8, 145, 193, 197, 212
    *(See also* Topology)
    fieldbus, 213 - 215
    serial, 217
    token, 214
Byte, 4

C programming language, 152, 241
Calibration
   temperature, 147
Chemical sensor
   array, 71 - 72
   biosensor, 77
   CHEMFET, 73
   chemicapacitor, 71 - 72
   chemiresistor, 71 - 72
   chromatography, 74 - 75
   humidity, 72, 76 - 77
   ISFET, 73 - 74
   polarography, 75
   potentiostat, 75
Circuit
   printed (PCB), 122 - 125, 198
   series tuned, 31 - 32
CMRR (common mode rejection ratio), 94
Coding
   efficiency, 5, 206 - 208
   frequency encoding, 218 - 219
   redundant, 145
Combustion, 195
Compensation, 143 - 144, 199
   classes of, 40 - 42, 198, 216
   deductive, 41 - 42
   frequency, 200
   monitored, 41, 198, 219, 225, 227 - 229, 232
   sensor-within-a-sensor, 41, 151
   structural, 40, 88, 198, 218 - 219, 221 - 223, 227, 232
   tailored, 41, 217, 219, 223
Constant
   Boltzmann, 70, 97, 110
   Plank, 83
   Verdet, 141
Control
   excitation, 143 - 144, 229
Convolution, 17 - 18, 25, 174
   integral, 17
Correlation, 114 - 115, 156
Costs
   fixed, 198
   variable, 198
Cross-correlation, 20, 65, 149, 200 - 201
Cross-sensitivity, 39, 146, 178, 217, 231
Cyclic voltammogram, 76
Czochralski process, 126


DAC, 100 - 102, 104, 120, 146, 151
   binary weighted ladder, 100
   R-2R ladder, 101 - 102
Damping, 30, 56, 195, 225, 227, 243
   ratio, 57
Data
   condensation, 6 - 7, 145
   conversion, 100 - 110, 143 - 144
Deconvolution, 17 - 18, 25, 173
Density
   signal, 109
Detector, 34
   box-car, 38
   ion, 192
   people, 192
   phase sensitive, 229
DFT, 157, 169 - 171, 176
Diaphragm, 44 - 46, 189, 228
Differentiation, 13, 119, 163 - 164
Digital filters
   *(See* Filter, digital)
Digital-to-analogue converter
   *(See* DAC)
Discrete Fourier Transform
   *(See* DFT)
Displacement sensor
   capacitive, 50 - 52
   inductive, 52 - 54
   optical, 54
   resistive, 46 - 50
Distribution, 19
Domain
   frequency, 13
   frequency, complex, 182
   $s$, 25

# INDEX

time, 13
$z$, 25, 161
Drift, 221, 225, 227
   parameter, 39 - 40

EEG, 175, 177
Effect
   Doppler, 20, 65, 175 - 176
   Faraday magneto-optic, 140
   Hall, 78 - 82, 217
   magnetoresistive, 78, 80 - 81
   Peltier, 66, 77
   photoconductive, 84
   photoemissive, 84
   photovoltaic, 83
   piezoelectric, 58 - 59
   piezoresistive, 49
   pyroelectric, 85
   Seebeck, 66
   thermoelectric, 66
   thermoresistive, 68 - 69
   Thomson, 66
   triboelectric, 112
Electrochemical cell, 75
Electromagnetic spectrum, 82 - 83
Entropy, 5, 206 - 207
Epitaxial growth, 126
Equation
   continuity, 7
   difference, 12, 31
   differential, 12, 30, 182
   diffusion, 128
   discrete, 30
   Doppler, 175
Ergodicity, 19, 22

Fast Fourier Transform
   (*See* FFT)
Fault, 212
   location, 197, 205
   tolerance, 197
FFT, 169, 171

Film
   Langmuir-Blodgett, 134
   thick (*See* Thick-film)
   thin (*See* Thin-film)
Filter, 17, 27, 113
   adaptive, 182 - 183, 185,
      187 - 188, 225, 227, 243
   analogue, 113, 143 - 144
   band-pass, 113
   Butterworth, 161 - 162
   Chebychev, 161
   design, 160
   digital, 37, 153, 156, 159 - 160,
      162 - 163, 199
   elliptic, 161
   high-pass, 113, 159
   low-pass, 39, 42, 113, 159, 180,
      202
   matched, 174, 177
   non-linear, 184
   non-recursive, 28, 160
   notch, 113
   recursive, 157, 160
   smoothing, 185
Fire, 192
Flow
   mass flow rate, 59
   variable area devices, 62
   velocity, 59
   volumetric flow rate, 59
Flowmeter
   Coriolis, 59
   cross-correlation, 65
   differential pressure, 59 - 61
   electromagnetic, 63
   hot-wire anemometer, 36,
      63 - 64
   laser Doppler, 65
   orifice plate, 59
   Pitot tube, 59, 62
   turbine, 63
   ultrasonic, 65
   variable-area, 62
   Venturi tube, 59, 61
   vortex-shedding, 63

Force
  Lorentz, 78, 80
Force sensor, 54
  (*See also* Load cell)
  cantilever beam, 55
  coupled-double-beam, 56
Function, 10, 223
  block, 28, 173, 202
  comb, 23
  cost, 181
  delta, 23
  density, 2, 19
  error, 198
  impulse, 12, 172, 191, 200
  *Si,* 203
  *sinc,* 28, 172 - 173, 203
  sinusoidal, 12, 200
  step, 13, 191, 200 - 201
  stimulus, 200
  system, 15
  transfer, 160
Fuzzy logic, 234

Gain, 19, 149, 223
  control, 109
GPIB, 211
Ground loops, 112

HART, 214 - 215
HDLC, 208 - 209, 220
Hold, 24
Humidity, 191
  dew point, 76
  psychrometric sensor, 77
Hygrometry, 76
Hysteresis, 178

Ignition, 195
Information
  definition, 4 - 6
  flow, 6 - 8
Instability

electrohydrodynamic, 201
Integration, 119, 163, 166
Interfacing
  circuits, 88 - 92
  low-power, 92 - 93
Interference, 112, 227
Interferometer, 138
Ion implantation, 128


Katharometer, 74


Law
  Faraday, 63
  Kirchhoff, 12
  Newton's second, 56
  Paschen, 51, 130
Line
  monitoring, 146, 228 - 229
Linearity, 11, 217
Linearization, 153
Load
  eccentricity, 222, 224
Load cell, 41, 179 - 180, 182 - 185,
    187 - 189, 198, 216,
    220 - 225, 227
  intelligent, 220 - 227
  tri-beam, 224
Look Up Table (LUT), 37, 153 -
    155, 179, 193, 195, 199,
    203, 219
LSB (Least Significant Bit), 2, 104,
    121
LVDT, 44, 52, 54, 59


Magnetic sensor, 217 - 219
  barber pole arrangement, 80
  Corbino disk, 81
  Hall effect, 78 - 79
  magnetoresistor, 80 - 81, 137
  magnetotransistor, 81, 217
MAP, 212 - 213
Mean

definition, 19
running, 27, 145, 163
square, 19
Measurement
definition, 1 - 3
Mechanical sensor
acceleration,
(*See* Accelerometer)
displacement,
(*See* Displacement sensor
flow,
(*See* Flowmeter)
force,
(*See* Force sensor)
pressure,
(*See* Pressure sensor)
strain gauge, 49 - 50
Microdynamics, 130
Microengineering, 228
Micromachining
silicon, 129 - 131
Microprocessor, 27, 146, 180
MOS, 125, 217
MSB (most significant bit), 104


Neural nets, 234
Noise, 38 - 39, 207, 212, 225
$1/f$, 39, 74, 111, 119
amplification, 118
digital, 114
quantization, 115 - 120
shot, 111
thermal, 110 - 111
white, 110 - 111, 174
Non-linearity, 37, 153, 180
bridge, 91
Non-stationarity, 20, 201
Nose
electronic, 217, 231 - 235
mammalian, 179, 232
Number, 3
Reynold's, 60

Offset printing, 123
Op-amp
(*See* Amplifier, operational)
Optical fibre, 137, 216, 227
Optical fibre sensor
extrinsic, 138
frequency modulation, 140
hybrid, 92 - 93
intensity modulation, 138
intrinsic, 138
microbend, 138
phase modulation, 138
polarization modulation, 140
Optimization, 157 - 158, 180, 227
Orthogonality, 167
Output
analogue, 146


Parity, 6, 145
Pattern recognition, 232
Petit mal, 176
Photolithography, 127
PIN diodes, 86
Plane
complex frequency, 13
half, 16, 26
$s$, 16 - 17, 161
$z$, 161 - 162
Poisson's ratio, 46, 48 - 49
Polarography, 75, 137
Pole, 15 - 18, 26, 30, 36 - 37, 163, 174, 181 - 182, 184 - 185, 187, 225
Polling, 145
Polynomial, 14, 24 - 25
$s$, 13, 15
$z$, 25
PRBS, 200
Precision, 2
Pressure
absolute, 42
atmospheric, 42
gauge, 42
Pressure sensor, 42, 227 - 229

Bourdon tube, 43
diaphragm, 44 - 46
U-tube manometer, 42
Principal component analysis
(PCA), 233
Principle
uncertainty, 29, 172
Priority, 205
Process
block, 157, 159, 232
evolutionary, 22
non-linear, 221
sampling, 23
stationary, 19
stochastic, 18
stream, 157, 159
Processing
communication, 143, 145
information, 143 - 144, 219, 225, 229, 232
Processor
control, 238
million transistor, 235
roving slave, 237
Protocol, 207 - 209, 214
master/slave, 214
PTAT, 70, 225

Q factor, 32, 228
Quantization, 22, 37, 109, 115 - 116, 118

Radiant sensor
bolometric, 85
CCD, 86
photoconductive, 84
photodiode, 85
photoemissive, 84
phototransistor, 85 - 86
photovoltaic, 83 - 84, 137
pyroelectric, 85
Redundancy, 5, 208
Relative humidity, 76

Response
frequency, 12, 17, 36, 174, 201
impulse, 12, 20, 201
time, 36, 201
RMS value, 22
ROM, 217, 220
Root loci, 31 - 32, 187
Rounding, 118, 120 - 121

Sample-and-hold, 102-103, 105
Screen printing,, 123 - 124, 135 - 137
Self-calibration, 241
Self-check, 148, 220
Self-test, 196, 241
Sensing element
primary, 143
Sensor, 33
array, 71 - 72, 137, 151, 177, 231
automotive, 193
chemical,
(*See* Chemical sensor)
defects, 36 - 40, 42
definition, 33
fuel, 193
gas, 195
knocking, 195
magnetic,
(*See* Magnetic sensor)
mechanical
(*See* Mechanical sensor)
micromachined, 44, 77
network, 204
optical fibre
(*See* Optical fibre sensor)
people, 193
radiant,
(*See* Radiant sensor)
smoke, 192
stimulus-response, 200 - 203
sub-system, 35 - 36, 90
temperature,
(*See* Thermal sensor)
two-port model, 34 - 35
Series

# INDEX

Fourier, 167
time, 29
Signal
  definition, 10
  recovery, 112
Silicon planar technology, 125
  crystal growth, 126
  diffusion, 127
  epitaxial growth, 126
  etching, 127
  ion implantation, 128
  metallization, 128
  oxidation, 126
  photolithography, 127
Simplex, 181
Smoothing, 28, 166
Space
  classification, 232
  feature, 232
  $n$-dimensional, 179, 181
  pattern, 232
Specificity, 41, 231
Specifier, 3
Spectrum, 20
  power, 18
Sputtering
  DC, 133 - 134
  RF, 133 - 134
Stationarity, 19, 201
Strain, 45 - 46, 48 - 50, 55 - 56, 58, 222, 228 - 229
Strain gauge, 44, 47, 49, 55, 92, 228
  gauge factor, 49 - 50
  metal foil, 50
  semiconductor, 49
  thick-film, 50
Stress, 222, 228
Strip
  bi-metallic, 228
Structure
  general purpose, 151
  hardware, 146
  indirect software, 157, 200
  minimal, 146
  signal processing, 157

software, 152
Symmetry
  design, 41, 88, 218 - 219, 222, 227
System
  analogue, 215
  causal, 24
  classification, 10
  communication, 209, 211 - 214
  continuous, 9
  definition, 9 - 10
  digital, 24, 27, 214
  discrete, 9
  distributed, 8
  first order, 30 - 31
  industrial, 204
  linear, 11, 115, 200
  master/slave, 214
  multisensor, 204
  non-causal, 203
  open, 209, 213
  resonant, 32
  response, 200, 205
  second order, 30 - 32, 182, 195, 216, 222
  topologies, 204
  two-port, 34 - 35

Temperature, 65 - 66, 68 - 70, 91, 190, 192, 196, 219
  coefficient of resistance, 68
  sensor,
    (*See* Thermal sensor)
Theorem
  Bernoulli, 59
  convolution, 17, 169
  fundamental, of algebra, 14, 243
  Heavyside expansion, 16
  sampling, 26, 173, 242
  Thevenin, 88
Theory
  information, 4
Thermal oxidation, 126
Thermal sensor

diode, 69
resistance thermometer, 68, 92, 137
thermistor, 69
thermocouple, 66 - 67
thermopile, 68
thermoresistive, 68
transistor, 69
Thick-film, 134, 146, 198, 219, 224, 232, 241
   definition, 131
   process, 135 - 136
   sensor, 71 - 72, 134, 136 - 137
Thin-film, 232
   chemical vapour deposition (CVD), 132
   definition, 131
   evaporation, 131 - 132
   sputtering, 133
Time
   response, 221
Topology
   bus, 204 - 209
   star, 145, 204
Transducer
   definition, 33
Transform
   bilinear-$z$, 161
   Discrete Fourier (DFT), 168
   Fast Fourier (FFT), 169
   Fourier, 13, 17, 20, 28, 168, 180, 201 - 202
   Laplace, 13 - 14, 17, 23 - 24, 31, 164, 201
   linear, 233
   $z$, 24 - 25, 28, 31, 161, 183
Transformation
   discrete, 166
   frequency, 162
Transient recorder, 156, 175
Trimming, 198
Truncation, 118

Variable
   flux, 34, 190
   physical, 34
   potential, 34, 190
   random, 2
VCO, 218 - 219

Warping, 161
Waveform
   recognition and classification, 175
   triangular, 20 - 21
Weighing
   non-linearity of, 189, 227
Window problem, 29, 169, 172 - 173

Young's modulus, 48, 222 - 223

Zero, 15, 17 - 18, 30, 36 - 37, 163, 174, 181, 184, 201